PROTECTING EMERGENCY RESPONDERS

VOLUME 2

Community Views of Safety and Health Risks and Personal Protection Needs

Tom LaTourrette, D. J. Peterson, James T. Bartis,
Brian A. Jackson, Ari Houser

Prepared for the
National Institute for Occupati

RAN
SCIENCE AND TECHNOLOG

D1416374

The research described in this report was conducted by RAND's Science and Technology Policy Institute, under Contract ENG-9812731.

Library of Congress Cataloging-in-Publication Data

Protecting emergency responders : community views of safety and health risks and personal
  protection needs / Tom LaTourrette ... [et al.].
      p. cm.
    "MR-1646."
    Includes bibliographical references.
    ISBN 0-8330-3295-X (pbk.)
    1. Emergency medical personnel—United States—Safety measures. 2. Fire fighters—United
States—Safety measures. 3. Police—United States—Safety measures.
    [DNLM: 1. Emergency Medical Technicians. 2. Emergency Medicine. 3. Community
Networks. 4. Risk Assessment. 5. Safety. W 21.5 P967 2003] I. LaTourrette, Tom, 1963–

RA645.5.P76 2003
362.18—dc21

                                                                    2003010699

RAND is a nonprofit institution that helps improve policy and decisionmaking through research and analysis. RAND® is a registered trademark. RAND's publications do not necessarily reflect the opinions or policies of its research sponsors.

Published 2003 by RAND
1700 Main Street, P.O. Box 2138, Santa Monica, CA 90407-2138
1200 South Hayes Street, Arlington, VA 22202-5050
201 North Craig Street, Suite 202, Pittsburgh, PA 15213-1516
RAND URL: http://www.rand.org/
To order RAND documents or to obtain additional information, contact
Distribution Services: Telephone: (310) 451-7002; Fax: (310) 451-6915; Email:
order@rand.org

FOREWORD

The National Institute for Occupational Safety and Health (NIOSH) is very pleased to have made possible this report conveying community views of health and safety risks and the personal protective needs for emergency responders. These views of occupational hazards and personal protective needs, gathered from emergency responders, will play a central role in NIOSH's continuing efforts to better protect our nation's emergency responders though improved technology, education, and training.

NIOSH is the federal agency responsible for conducting research and making recommendations for the prevention of work-related disease and injury. Created by Congress in 1970 with the passage of the Occupational Safety and Health Act, the Institute is part of the Centers for Disease Control and Prevention within the Department of Health and Human Services. Its mission is to provide national and world leadership in preventing work-related illness, injury, and death by pursuing the strategic goals of surveillance, research, occupational disease and injury prevention, and information and training.

In fiscal year 2001, Congress allocated funds for NIOSH to establish a new program for personal protective technology research to protect the nation's miners, firefighters and other emergency responders, and health care, agricultural, and industrial workers. To carry out this research, NIOSH formed the National Personal Protective Technology Laboratory (NPPTL). The Laboratory's mission, like the mission of its parent organization, is to provide world, national, and Institute leadership for prevention and reduction of occupational disease, injury, and death but with special emphasis on those workers who rely on personal protective technologies.

The NPPTL is engaged in an active program of research, standards development, and information dissemination. Recently, the Laboratory developed test methods and standards for self-contained breathing apparatus and gas masks that could be used in the event of a chemical, biological, radiological, or nuclear terrorist attack. The tragic events of September 11, 2001, underscore the significance of the mission of the NPPTL. The lessons learned from those events identify several important areas that warrant attention and are providing critical guidance for our research.

Richard Metzler
Director, National Personal Protective Technology Laboratory
National Institute for Occupational Safety and Health

Firefighters, law enforcement officers, and emergency medical personnel play a critical role in protecting people and property in the event of fires, natural disasters, medical emergencies, and actions by terrorists and other criminals. This report presents an overview of occupational hazards and personal protection needs as viewed by emergency responders in the United States.

The primary goal of this report is to help define technology needs and research priorities for personal protection for emergency responders. Feedback from expert stakeholders is essential to this process. The findings reported here were derived from discussions with 190 representatives from 83 organizations in the emergency response community nationwide. These findings are intended for use in conjunction with emergency responder injury and fatality data, evaluations of current personal protection research, and assessments of existing personal protective technologies to help federal managers and decisionmakers to

- understand the evolving work and safety environment surrounding emergency situations

- develop a comprehensive personal protective technology research agenda

- improve federal education, training, and other programs directed at the health and safety of emergency responders.

This report was requested by the National Personal Protective Technology Laboratory of the National Institute for Occupational Safety and Health. The Laboratory was created in 2001 to ensure that the development of personal protective equipment keeps pace with employer and worker needs as work settings and worker populations change and new technologies emerge. The Laboratory's initial area of emphasis is to respond to the critical need for effective personal protective technologies for the nation's emergency responders.

This report should be of interest to agencies involved in research, implementation, and guidance associated with protecting emergency responders. This report should also help state and municipal officials, trade union leaders, industry executives, and researchers to gain a better understanding of the various equipment and training needs for protecting emergency workers.

This report is the second in a series of RAND publications on *Protecting Emergency Responders*. The first in the series is

- Brian A. Jackson, et al., *Protecting Emergency Responders: Lessons Learned from Terrorist Attacks,* CF-176-OSTP, 2002 (available at http://www.rand. org/publications/CF/CF176/).

The study approach and findings in this report also build on the following earlier RAND studies on related areas of research:

- William Schwabe, Lois M. Davis, and Brian A. Jackson, *Challenges and Choices for Crime-Fighting Technology: Federal Support of State and Local Law Enforcement,* MR-1349-OSTP/NIJ, 2001 (available at http://www.rand. org/publications/MR/MR1349/)

- D. J. Peterson, Tom LaTourrette, and James T. Bartis, *New Forces at Work in Mining: Industry Views of Critical Technologies,* MR-1324-OSTP, 2001 (available at http://www.rand.org/publications/MR/MR1324/).

THE SCIENCE AND TECHNOLOGY POLICY INSTITUTE

Originally created by Congress in 1991 as the Critical Technologies Institute and renamed in 1998, the Science and Technology Policy Institute is a federally funded research and development center sponsored by the National Science Foundation and managed by RAND. The Institute's mission is to help improve public policy by conducting objective, independent research and analysis on policy issues that involve science and technology. To this end, the Institute

- supports the Office of Science and Technology Policy and other Executive Branch agencies, offices, and councils

- helps science and technology decisionmakers understand the likely consequences of their decisions and choose among alternative policies

- helps improve understanding in both the public and private sectors of the ways in which science and technology can better serve national objectives.

Science and Technology Policy Institute research focuses on problems of science and technology policy that involve multiple agencies. In carrying out its mission, the Institute consults broadly with representatives from private industry, institutions of higher education, and other nonprofit institutions.

Inquiries regarding the Science and Technology Policy Institute may be directed to:

Helga Rippen
Director, RAND Science and Technology Policy Institute
1200 South Hayes Street
Arlington, VA 22202-5050
Phone: (703) 413-1100 x5574
Web: http://www.rand.org/scitech/stpi/
Email: stpi@rand.org

CONTENTS

FIGURES

Emergency response is an inherently dangerous occupation. Emergency responders face a wide range of serious hazards in their jobs, which places them at high risk for occupational injury or death. This risk is mitigated by their using various forms of personal protective technologies (PPTs), such as protective garments, respiratory protection, environmental monitoring and communications equipment, and practices and protocols that focus on safety.

This report addresses the safety of emergency responders by examining the hazards and personal protection needs that members of the emergency responder community regard as being the most important. The findings reported here are based on in-depth discussions with 190 members of the emergency response community nationwide, including structural firefighters, emergency medical service (EMS) responders, police officers, emergency management officials, technology and services suppliers, researchers, and program managers from 83 organizations around the country.

The principal topics addressed in this report include:

- The primary tasks that emergency responders undertake
- Situations in which the risk of injury is the greatest and that have the highest priority for improving personal protection
- Current and emerging technologies that are critical to protecting the health and safety of emergency responders
- Drivers of, impediments to, and gaps in technology development.

PROTECTING FIREFIGHTERS

Firefighters who participated in this study consistently noted that their protective clothing (turnouts or bunker gear)[1] provides excellent flame retardance and thermal protection. However, despite the high protective capability of current firefighter clothing materials and components, several protection challenges remain.

A firefighting ensemble composed of highly effective components can nevertheless leave firefighters vulnerable to injury due to component incompatibility or bodily exposure at component interfaces, with mismatched gloves and coat cuffs often cited as examples. To address such problems, study participants recommended increased "configuration control"—the standardized specification of component dimensions and interfaces.

Reducing thermal and physical stress is a top priority among the firefighters with whom we met. The thermal protective ensemble, including turnouts, boots, gloves, and hoods, almost completely encapsulates a firefighter, which creates difficulties in dissipating body heat. The weight of the protective garments, self-contained breathing apparatus (SCBA), and firefighting equipment puts firefighters at high risk of injury from physical stress and overexertion. Study participants pointed to several approaches to addressing this problem, including increasing the vapor transmission of turnout textiles and improving the fit of turnout gear to increase its flexibility and comfort. Another suggested approach is the implementation of physiological monitoring and communications systems to provide advance warning before firefighters suffer heat stress or exhaustion.

Firefighters noted that they are generally very satisfied with the respiratory protection afforded by modern SCBA. However, study participants also observed that there are situations in which alternative forms of respiratory protection may be appropriate, such as during fire overhauls[2] or during search-and-rescue operations after a structural collapse. Some participants cautioned, however, that any such alternatives would provide less respiratory protection, a consideration that must be weighed carefully in any decision. Discussion participants also called for ways to improve SCBAs, citing the desire for lighter and higher-capacity air bottles and improved air supply monitoring and warning capabilities.

---

[1]Firefighter protective clothing, commonly referred to as *turnouts* or *bunker gear,* consists of flame- and water-retardant pants and overcoat.

[2]*Fire overhaul* begins when the main fire has been suppressed. It entails activities such as searching for hidden hot spots, salvaging property, and cleaning up debris and equipment.

Improving communications for individual firefighters is another high-priority area mentioned by study participants. They repeatedly pointed out that firefighters have great difficulty communicating person-to-person and over a radio while wearing an SCBA face mask. Some participants further observed that their radios are not designed specifically for the needs of a firefighter, which is a result of the relatively small market share that emergency responders represent.

Improving *fireground accountability,* the ability to account for the whereabouts of firefighters at an incident scene, was also viewed by larger fire departments as a high priority. Many firefighters are injured or do not receive prompt treatment for injuries, participants claimed, because of confusion over the location and activities of individuals during an incident. Existing accountability systems that rely on manually transferring personal identification tags to status boards were viewed as being outdated. Innovations utilizing magnetic card readers, which were discussed by several participants, may provide improved accounting system flexibility and reliability.

PROTECTING EMERGENCY MEDICAL SERVICE RESPONDERS

Discussion participants representing the emergency medical services commonly claimed that little protective equipment designed specifically for their work environment is available. And what does exist is often low quality, uncertified, or impractical. To remedy this problem, some organizations were adopting PPT, such as SCBAs, bunker gear, and armored vests, from the fire and law enforcement services. One reason cited for the shortfalls in EMS protection is that no federal agency is dedicated to addressing personal protection issues, such as equipment, standards development, certification, and PPT usage enforcement for the emergency medical responder community, and little funding is dedicated to address these issues. Addressing protection needs is further complicated by the wide range of tasks that EMS responders undertake and the multiple types of agencies that provide emergency medical response service.

Emergency medical service responders expressed a strong concern about exposure to infectious diseases such as AIDS, hepatitis C, and tuberculosis. Although exposure to infectious diseases accounts for very few actual responder injuries or illnesses, pathogens were seen as a growing hazard and one of the most difficult hazards to protect against. Emergency medical responders typically have access to protective gloves, masks, goggles, and splash gowns. However, this gear is often designed for hospital use and is sometimes difficult to use in the field. Study participants in several EMS departments noted that usage of this gear has increased considerably through the issuance of fanny packs containing an ensemble of protective gear, which make the gear more easily accessible.

Another hazard of increasing concern to EMS personnel is physical assault. Unpredictable circumstances leave EMS responders particularly vulnerable to surprise attacks and other violent acts. In response, many EMS personnel are now being trained in situation management and self-defense. EMS responders in many larger departments are also being issued body armor. However, the use of body armor is left to the discretion of individuals, and its use is estimated to be rare.

Like emergency responders in all services, EMS responders are concerned about hazards associated with terrorism. The top concern in this area is exposure to biological and chemical warfare agents, either direct exposure or exposure while treating victims. EMS participants expressed a desire for improved hazard assessment training, as well as better respiratory protection and protective clothing options, to deal with these hazards.

PROTECTING LAW ENFORCEMENT OFFICERS

A conclusion that emerged from our discussions with law enforcement representatives is that protecting law enforcement personnel may be the most challenging personal protection task within the emergency response community. This finding stems from several factors: Law enforcement responders are typically the first responders on the scene of an incident and hence have the least advance information about potential hazards; their mobility and patrol requirements limit the amount of gear they can wear or carry with them; their appearance requirements, particularly for covert operations, limit their protection options; their being on patrol rather than returning to a station between calls limits training opportunities; and most personal protective technologies are not developed with the law enforcement mission and operating environment in mind. In addition, law enforcement lacks a centralized professional organization dedicated to health, safety, and protection. As with EMS, law enforcement often turns to fire service resources for guidance.

The ballistic vest is the most widely used personal protection technology in law enforcement. Despite their proven effectiveness, police often do not wear vests because they can be hot and uncomfortable, particularly while riding in a car. Vest designs have improved over the years to address these concerns, but the design improvements have been achieved, in part, by reducing the size of vests, and some participants expressed concern that body coverage was too small. Alternatives such as "throw-on" armored jackets were mentioned as an option, though participants noted that those jackets might not be readily available when needed.

Automobile injuries are another area of concern. Representatives from a number of departments noted three main problems contributing to automobile driving hazards: (1) The side placement of computers and radios can cause officers to become distracted while driving and can present impact hazards in accidents. Study participants called for in-dash systems and overhead displays to improve safety. (2) High-speed, rear-end collisions are also a serious problem, and participants suggested strengthening automobile frames, adding rear-impact safety devices, and improving vehicle warning lights. (3) Finally, unsafe driving behavior, particularly in younger officers, is a major contributor to accidents and could be mitigated by stricter driving policies or by speed monitoring or governing systems.

Pathogen protection is another concern among law enforcement responders, particularly protection from pathogens transmitted during physical assaults such as biting or spitting. While many patrol cars are stocked with disposable gloves and sometimes also masks, these items are difficult to access quickly and are rarely used.

PROTECTING RESPONDERS FROM TERRORISM

A concern expressed by the entire emergency responder community is adequate protection against terrorist attacks and the vulnerability of nonspecialist first responders in particular. Accordingly, several emergency responder departments have begun equipping their vehicles with chemical protective gloves, suits, escape hoods,[3] and respirators.

RAND's discussions with participants revealed that the issue of providing protection for chemical, biological, or radiological (CBR) terrorism is complicated by several uncertainties:

- Many police and fire department representatives felt that they did not know what they need to be protected against, what form of protection is appropriate, or where to look for such protection. Such uncertainty frustrates efforts to design a protection program and acquire the necessary technology.

- Participants were unsure how well the available protective technologies will work for anticipated situations. While hazardous materials (hazmat) protection is subject to rigorous standards and certification procedures, hazmat equipment and usage protocols are designed primarily around the conventional model of hazmat response to industrial accidents. Much of the available hazmat protection is neither designed nor certified for this new role of terrorism response.

---

[3]An *emergency escape hood* is a soft-sided pullover hood with an elastic neck seal. These hoods provide particulate and chemical respiratory protection to enable wearers to exit hazardous environments.

- Participants were unclear how personal protective technology is expected to be used in terrorist events. Because of the uncertainty surrounding the roles of responders in such situations, major questions remain as to exactly where such equipment should be stored, when it should be donned, what tasks should be performed while it is used, and who should make these decisions.

BEYOND THE INDIVIDUAL: SYSTEMS-LEVEL PROTECTION

In addition to protective clothing and other personal gear that supports a single individual, several other forms of emergency responder protection operate at the command or unit level. Such "systems-level" protection mentioned by participants includes communications, location monitoring, hazard monitoring, and various human factors.

Communications

Beyond the tactical communications issues that firefighters face (discussed above), a number of police, EMS, and fire departments emphasized strongly that there are fundamental problems with the radio communication systems currently used by emergency responders. Departments often use incompatible radio systems and cannot communicate easily with each other at the scene of major incidents. This problem affects communications among local departments as well as communications between municipal departments and state or federal agencies. Such communications breakdowns can have severe consequences. For example, incident commanders may have difficulty in maintaining scene control, utilizing forces most effectively, or sharing critical safety information.

This problem is being addressed by a push toward implementing a uniform, interoperable radio system for emergency responders. While this radio system—a digital, 800-megahertz backbone system—has many advantages over analog radio-to-radio technologies, many departments that had acquired these systems were not fully satisfied with their performance. Their concerns include the inability to talk over other users, unreliable signal transmission in areas with tall buildings or hills, and the high investment costs. As a result, departments often resort to maintaining multiple systems to handle all of their communications needs.

Hazard Assessment

An important part of protecting emergency responders is understanding the hazards that they face. While generalized models based on empirical evidence provide much of the basic input on protection choices, incident-specific information can further characterize those hazards and inform protection and procedural decisions. Several hazard-assessment tools were mentioned in the discussions, including:

- On-site information, such as hazmat placards
- Facility "pre-plans"[4]
- Information supplied by dispatchers
- Environmental monitoring equipment.

Participants noted that all of these methods can provide useful information, but that they suffer from various shortcomings that limit their applicability. Interestingly, most participants stated that hazard information is often used to guide operational decisions but rarely influences personal protection selection because protection options are very limited to begin with.

Personnel Location Monitoring

A longer-term but potentially very valuable technology for larger services is personnel location monitoring. Participants from both fire and police departments made mention of this technology and noted that the primary benefit would be the ability to quickly locate a trapped or injured responder. The technology could also assist in managing operations, guiding personnel through buildings, improving dispatching efficiency, and managing driving behaviors. Several participants have begun investigating emerging technologies based on the Global Positioning System (GPS). Such systems, however, are expensive and, more fundamentally, suffer from poor vertical resolution and signal penetration problems. Other location technologies under discussion and in development utilize radio triangulation (exploiting differences in travel times of radio signals between a source and multiple receivers), radar (exploiting the travel time of reflected radio signals), inertial tracking (using accelerometers to compute cumulative movement; also known as "dead-reckoning" systems), and hybrid systems.

---

[4]*Pre-plans* comprise site-specific information compiled beforehand, such as information on hydrant and standpipe locations, utilities, building design and layout, hazardous material inventories, and service histories from previous calls.

Human Factors

Human factors play an important role in emergency responder safety and health. As data collection and manipulation capabilities increase, limitations in knowledge management, or the ability of people to effectively utilize available information, can impact responder safety in some cases. Commonly cited examples include underutilization of mobile data terminals and the inability to use or correctly interpret readings from environmental hazard monitors.

Another critical human factor is adoption of safety practices to mitigate day-to-day injuries, such as a sprain from a fall. Several agencies are addressing these hazards with standard approaches such as offering physical fitness classes, maintaining a safe environment in fire stations, and issuing properly fitting clothing and supportive footwear.

Tradition and culture also affect emergency responder safety. A common example is a preference for a certain style of fire helmet: Despite their substantial weight and higher cost, many firefighters prefer the appearance of traditional-style helmets with large brims. Another cultural aspect that may impact safety is the fraternal and often voluntary nature of the profession, which can temper enforcement of safety practices. In this regard, many participants pointed to the more stringent standards used by specialized units such as hazmat or urban search-and-rescue teams. Finally, tradition may hinder the adoption of safety and health innovations. Decisions on whether to accept new technologies or even simply to change brands or suppliers are deeply rooted in tradition.

PROCUREMENT AND LOGISTICS OF PROTECTIVE TECHNOLOGIES

Decisions on how PPTs are identified, acquired, and used in the field vary significantly, as was noted by many participants. Many issues and concerns were raised on the procurement and logistics of protective technologies that have implications for PPT research and development needs.

Personal Protective Technology Standards and Performance Evaluation

A critical concern for most departments was their getting adequate information to guide technology acquisitions. Participants indicated that few emergency response agencies have the resources or capabilities to conduct formal risk assessments to guide these acquisitions. As such, many departments choose protective technology based on supplier relationships. While design and performance standards assure a basic level of functionality and protection, distinguishing among the large variety of certified gear within each equipment class is not a straightforward process.

Consequently, most responder organizations must resort to informal, ad hoc PPT evaluation and information gathering and analysis because they lack access to reliable public sources on PPT performance that would inform their procurement decisions. In response to these problems, many participants strongly advocated implementing objective, third-party assessments to help guide them in their PPT evaluations and decisionmaking.

Storage and Maintenance

As emergency responders have acquired greater amounts of protective equipment, storerooms, vehicles, and people have become increasingly crowded and burdened. An individual can carry only so much gear. Squad car trunks are getting full. EMS vehicles have limited storage space. Many communities have purchased dedicated disaster response vehicles or trailers, and many have created supplemental equipment caches, but these measures raise questions about how rapidly such equipment will be fielded and who will have access to it.

As emergency response organizations acquire greater amounts of gear, their equipment maintenance and reliability needs are also increasing. Many emergency responders mentioned the strain that meeting these needs places on a department. Firefighters expressed concern over their departments' ability to ensure the integrity of turnouts (moisture barriers in particular) and other gear after extensive use. Several fire and police departments as well as PPT manufacturers felt that passive integrity monitors, such as indicators that change color as material properties change, would be a valuable addition to protective equipment. Along with the availability of sophisticated environmental monitoring and other electronic equipment comes the need for technical expertise and resources to maintain that equipment.

Universal Versus Tailored Personal Protective Technology

The role of emergency responders continues to expand as does the ability of emergency responders to evaluate site-specific hazards. Thus, several participants claimed, opportunities exist to improve safety by selecting protection options that are based on the specific situation. However, such options are currently quite limited.

The standard in the fire service is universal protection—a single ensemble designed to protect against all anticipated hazards. Such an ensemble is optimized for structural fires and may not provide the best protection for the range of other situations firefighters encounter, such as vehicle accidents or medical calls.

Arguments against tailored protection include the simplicity that a single ensemble affords, uncertainties about the actual hazards, and the time, cost, and energy involved in supporting several types of protective clothing. Risk-specific protection is beginning to emerge: Protective clothing standards for urban search-and-rescue and emergency medical response ensembles recently have been introduced.

Interoperability

A final logistics issue concerns mutual aid[5] agreements between jurisdictions and the interoperability of protective equipment. Interoperability of protective equipment may be critical at large incidents, as was the case with respirators at the World Trade Center in September 2001. Mutual aid agreements between jurisdictions typically address incident management, training, and technical capabilities, but protection is rarely included in this list. Major barriers to PPT coordination in the emergency responder community include incompatibilities in funding cycles, equipment replacement cycles, and purchasing power; tradition and well-established vendor relationships that hinder change; and the absence of procedures for accomplishing PPT coordination easily.

PUTTING COMMUNITY VIEWS TO WORK

A number of issues emerged from RAND's discussions with participants that have important implications for improving the protection of emergency responders. These issues generally can be divided between two areas: (1) priority areas for improving equipment and practices and (2) broader policy issues that warrant further research, analysis, and discussion. The priority areas are relatively straightforward and are, for the most part, consensus concerns within the responder community that were raised directly by the discussion participants. Many of the policy issues, on the other hand, are complex and pose challenging questions. These issues emerged indirectly from the community discussions, and most are marked by fundamental differences of opinion within the community. These issues are summarized in Tables S.1 and S.2. In several cases, these concerns are actively being addressed by government agencies and other organizations concerned with emergency responder safety.

---

[5]A *mutual aid* response is one in which more than one department participates.

Table S.1

Personal Protection Priorities and Recommendations Raised by the Emergency Responder Community

Personal Protection Priorities	Specific Recommendations
Reduce physical stress and improve comfort	• Improve garment breathability • Reduce equipment weight • Ensure consistent and appropriate sizing of components • Enhance ergonomic characteristics
Improve communications	• Make radio systems interoperable • Improve communications capabilities with SCBA • Improve radio design to allow hands-free use and use with gloves
Upgrade communicable disease protection	• Increase protective equipment options for EMS personnel and police
Develop practical respiratory and chemical protection equipment and guidelines for first responders	• Improve the chemical and biological protection of garments and respirators • Design protective equipment such that it minimizes interference with responder activities • Require more chemical/biological hazard training
Improve PPT standby performance	• Develop integrity monitoring and service-life monitoring technologies • Enhance compactness and portability of protective equipment • Address logistical complications • Reduce protective equipment maintenance complexity and cost
Expand training and education	• Require more training on sophisticated protective equipment • Reduce complexity of new equipment
Benchmark best safety practices	• Study and benchmark safety practices, particularly for EMS and police • Study and benchmark PPT enforcement practices

Table S.2

Key Policy Areas and Issues Raised by the Emergency Responder Community

Policy Areas	Specific Issues
PPT research and development	• Research should be more strategic and multidimensional, including more fundamental, long-term research • Greater emphasis on ensembles is needed • R&D should address response activity rather than services • Decentralized market limiting innovation and purchasing power should be addressed
Discretion in personal protection decisionmaking	• Expanding role of emergency responders and improved hazard assessment warrant increased attention to activity-specific tailoring of protection
PPT standards for emergency medical services and law enforcement	• EMS and police communities need dedicated personal protection, safety, and standardization efforts
PPT performance assessment	• Reliable and objective equipment performance assessments need to be developed
PPT standardization and interoperability	• Mutual-aid agreements and extended operations should be facilitated by enhanced standardization and interoperability
The role of risk in emergency response	• Examine emergency responders' perceptions of and their responses to risks inherent in emergency response • Promote efforts to decrease risk through improved information management, clarified protocols, and improved equipment

ACKNOWLEDGMENTS

We gratefully acknowledge the many members of the emergency responder community throughout the country who participated in the discussions and thank them for their time, thoughtfulness, and candor. A list of participants can be found in Appendix A.

We thank Richard Metzler, Louis Smith, and Jonathan Szalajda at the National Personal Protective Technology Laboratory for their guidance throughout the study. We also thank the following people for their assistance in identifying departments and other organizations to include in the study: William Haskell of the U.S. Army Soldier and Biological Chemical Command's Soldier Systems Center and the InterAgency Board for Equipment Standardization and InterOperability; Don Rosenblatt, executive director of the International Association of Chiefs of Police; Assistant Chief James Hone of the Santa Monica Fire Department; Kathleen Higgins of the National Institute of Standards and Technology Office of Law Enforcement Standards; Jim Gass of the Oklahoma City Memorial Institute for the Prevention of Terrorism; and Andy Levinson of the International Association of Firefighters. At RAND, we thank Jerry Sollinger for his help in preparing this report.

Finally, three peer reviewers provided important analytical insights and background information that strengthened this report: K. Jack Riley, director of RAND Public Safety and Justice; Robert C. Dubé, captain, Fairfax County Fire and Rescue; and Paul M. Maniscalco, executive council member of the National Association of Emergency Medical Technicians.

APR	Air-purifying respirator
BDU	Battle dress uniform
CBR	Chemical, biological, radiological
EMS	Emergency medical service
EMT	Emergency medical technician
FBI	Federal Bureau of Investigation
GPS	Global Positioning System
Hazmat	Hazardous material
HIV	Human immunodeficiency virus
ICS	Incident Command System
ISO	International Standards Organization
MHz	Megahertz
NATO	North Atlantic Treaty Organization
NFPA	National Fire Protection Association
NIJ	National Institute of Justice
NIOSH	National Institute for Occupational Safety and Health
NPPTL	National Personal Protective Technology Laboratory
OSHA	Occupational Safety and Health Administration
PPE	Personal protective equipment

PPT   Personal protective technology

R&D   Research and development

SCBA   Self-contained breathing apparatus

SWAT   Special weapons and tactics

USAR   Urban search and rescue

WMD   Weapons of mass destruction

Chapter One

INTRODUCTION

Every day in the United States, emergency responders answer calls for help and take on jobs that place them in harm's way. Firefighters, law enforcement officers, emergency medical technicians, and paramedics play a critical role in protecting the American public and property in the event of fires, natural disasters, medical emergencies, or actions by terrorists or other criminals. Emergency responders' responsibilities extend from dealing with small-scale, "everyday" emergencies that may affect only a single individual, family, or business, to responding to large-scale disasters such as earthquakes, hurricanes, or terrorist attacks. Accordingly, it is in the nation's interest to aid in the protection of these workers both for their own sake and to sustain their ability to protect the country.

STUDY TASK AND PURPOSE

In an effort to understand the range of hazards to which emergency responders are exposed and to identify critical protective technology needs, RAND conducted a series of structured, in-depth discussions with a wide range of representatives from the emergency response community. The study was requested by the National Personal Protective Technology Laboratory (NPPTL) within the National Institute for Occupational Safety and Health (NIOSH) to help guide development of a research agenda. NPPTL was created in 2001 with the mission of "providing world, national and [NIOSH] leadership for the prevention and reduction of occupational disease, injury, and death for workers who rely on personal protective technologies—through partnership, research, service, and communication." An important objective of NPPTL is to ensure that the development of personal protective technology (PPT) keeps pace with employer and worker needs as work settings and worker populations change and as new technologies emerge. NPPTL's initial area of emphasis is to respond to the critical need for effective personal protective technologies for the nation's emergency responders.

Among NPPTL's strategic goals are understanding the hazards for which personal protective technologies are used, the use and limitations of personal protective technologies and the programs guiding their use, the barriers to effective use of protective technologies, and personal protective technology failures. Accordingly, the objective of RAND's discussions with representatives from the nation's emergency responder community was to elicit the community's views on the following questions:

- What are the principal tasks that emergency responders undertake and how might those tasks change in the future?

- What are the occupational risks and hazards that are of greatest concern to emergency responders, and which are the highest-priority for improving protection?

- What are the current and emerging technologies critical to protecting emergency responders' health and safety and enhancing their capabilities?

- What are the drivers of, impediments to, and gaps (i.e., shortfalls in equipment availability, price, utilization, performance, or management) in technology development that limit progress in reducing injuries to and improving the capabilities of the emergency response workforce?

While a substantial amount of data is collected about emergency response activities and responder injuries and deaths, first-hand views from the emergency responder community about the hazards they face and their protection needs provide insights about those needs that cannot be derived from injury and fatality statistics. One significant problem with the data is that they provide no insight into the details on why different personal protection options do or do not work well. In addition, these data provide little information to link specific types of activities to specific types of injuries. The community views can reveal many important gaps in personal protective needs that are not apparent through analysis of available occupational health and safety surveillance data.

HOW THE STUDY WAS CONDUCTED

RAND researchers led structured discussions with 190 representatives from 83 organizations across a broad spectrum of the emergency responder community nationwide. The findings presented in this report are drawn largely from input provided by representatives from 61 local (i.e., front-line) emergency responder organizations. These organizations include 33 fire departments (28 city and county agencies, 4 private industrial and municipal services firms, and 1 volunteer department); 22 law enforcement agencies (19 city police departments, 1 state police department, 1 county sheriff's department, and 1 tribal police department); 3 independent (third-service)[1] emergency medical service (EMS) organizations; and 3 local emergency management offices.

These local agencies typically have responsibility for responding to structural fires, medical emergencies, transportation accidents, crimes, public disturbances, natural disasters, and terrorist acts within their jurisdictions. In arranging the discussions with representatives of the responder community, participants with special expertise in emergency medical services, hazardous material (hazmat), and special law enforcement operations response were specifically invited and participated in most discussions with departments that provided such special services.[2] Hazmat response is also handled primarily by fire departments, with 25 of the 33 participating fire departments providing hazmat response.

The rank of participants was mostly at the assistant, deputy, or battalion chief level, though several department chiefs as well as lower ranks were represented. Most participants had expertise in either special operations or protective equipment acquisition and maintenance. In some cases, safety officers, training officers, and occupational health experts participated in the discussions.

Participating organizations were selected according to several criteria. The goal was to sample a range in department size and type, socioeconomic composition of the community served, and geographic location. Departments were also included based on their reputation within the profession and on recommendations from other discussion participants. Finally, some departments were selected based on logistical considerations as well (i.e., how easily they could be accessed). In the end, the RAND sample was biased toward larger departments relative to the national average. This bias was intentional and was prompted by the understanding that larger departments generally have greater resources and capabilities for analyzing risks and assessing personal protective technology needs.

To supplement the information gathered from local agency representatives, the RAND team also met with representatives from 22 business, government, nongovernmental, and academic organizations, including 9 technology and services suppliers and 13 agencies and organizations engaged in PPT research, policymaking, and program development.[3] These community representatives,

---

[1]Service from a third party, after fire and police departments.

[2]Although only three independent EMS providers were included in the RAND roster, nearly all of the fire departments that were contacted are the primary EMS providers for their jurisdictions, and representatives specializing in EMS participated in most discussions. However, the EMS input in this study is biased toward the fire service and does not reflect the diversity of EMS delivery models that are in use.

[3]Due to logistical constraints, a few participants were contacted by telephone.

typically operating at the national level or focusing on a specific topic, provided the RAND team with important background information on technical issues related to personal protection and emergency response, research agendas, and programs and policies. Many representatives also assisted us by identifying local agencies for inclusion in the RAND discussions.

All of the discussions were conducted between March and July 2002. They typically were conducted on the participants' premises and lasted from 90 minutes to two hours; all discussions were held on a not-for-attribution basis.

Figure 1.1 shows the geographical distribution and type of organizations that participated in the study. Many organizations elected to delegate more than one representative to this research effort, bringing the total number of individuals who participated in the RAND study to 190. A list of discussion participants and their affiliations is provided in Appendix A. In almost every instance, we found the participants to be highly engaged, thoughtful, and willing to address even sensitive issues.

Figure 1.1—Location and Type of Participating Organizations

To maintain consistency, discussions were guided by a 20-question protocol developed by RAND in conjunction with NPPTL.[4] The protocol is reproduced in Appendix B. The protocol was designed to encourage participants to think broadly and creatively and pursue issues of special interest related to their particular localities or individual experiences, yet at the same time keep the discussion focused on the questions listed earlier in this chapter.

Most discussions were conducted by one or two members of a five-member RAND team. To further minimize inconsistency among discussions and to facilitate consistent interpretation of responses, 65 of the 83 discussions (78 percent)[5] were conducted by one or both of two team members, with both of those members present during 11 discussions (13 percent). All discussion notes were shared among team members, and team meetings were regularly held so that team members could share the input they received.

LIMITATIONS OF THE STUDY APPROACH

Our approach, which utilizes structured discussions to elicit the views and priorities of the emergency responder community, offers unique insights that are relevant to the questions surrounding the hazards and protection needs that emergency responders face. However, such an approach has significant limitations, particularly within the context of using the views of the community to inform the higher-level objective of defining research and development (R&D) priorities for personal protective technologies. While we have taken steps to mitigate these limitations, they nonetheless must be kept in mind when interpreting our findings.

One limitation is the qualitative nature of the input. Because of the broad scope of the discussion protocol, the wide range of types of agencies and organizations included in the study, and the individual nature of the discussions, the information collected from the discussions covered a vast range of topics and typically could not be quantitatively classified in certain ways, such as according to the exact number of participants or departments expressing a particular view. In addition, discussion participants were asked to express their personal views and did so with varying degrees of clarity and emphasis, which leaves open the possibility of inconsistency in interpreting responses.

---

[4]The protocol was tailored primarily for discussions with the 61 emergency responder departments. Discussions with the 22 business, government, nongovernmental, and academic organizations loosely followed the protocol, but tended to focus more narrowly on the organization's specific area of expertise.

[5]Of the 70 discussions with emergency responder departments and manufacturers and service providers, 62 of the discussions (89 percent) were conducted by one or both of two team members.

Another related limitation is the degree of reproducibility of the findings. Despite our use of a discussion protocol and our efforts to maximize consistency among RAND discussion leaders, each discussion was unique and depended on the roles and experience of the individual participants. We attempted to mitigate this effect by sampling a large number and wide range of organizations. This approach generally was successful in that clear themes emerged, and there were clear distinctions among issues with low, moderate, or high degrees of consensus. However, it is possible that a different sample or discussion approach would have yielded somewhat different findings.

A final limitation is the inherent incompleteness of and bias in the information that can be obtained solely from the viewpoint of the emergency responder community. Emergency responder organizations in the United States are very decentralized, and many agencies, particularly the smaller ones, may not be aware of certain initiatives or resources that are available to address various problems. As municipal agencies, emergency responder departments' budgets are often tight, and in many cases, the primary concern of emergency responders is not the availability of technologies but the availability of funds to acquire those technologies.[6] In addition, as end-users, many emergency responders are primarily interested in and knowledgeable of fully developed, tested, and accepted technologies. As a result, they may pay little attention to, or may even actively dismiss, some emerging technologies that are not fully developed or widely diffused. In so doing, they may misconstrue some of the community's needs. We attempted to balance this potential shortcoming by including in this report discussions of emergency responder injury and fatality data as well as descriptions of existing technologies, standards, and programs whenever they were relevant to concerns raised by participants.

DEFINITIONS

In this study, we adopted a broad definition of *technology*—namely, the application of knowledge toward practical ends. Accordingly, personal protective technologies include not only conventional protective equipment, such as clothing, gloves, respirators, and helmets, but also other physical hardware (e.g., detectors and communications systems) in addition to operational procedures, organizational structures, and management practices. The inclusive nature of this definition is important: According to the community members with whom we spoke, some of the most effective means for protecting emergency responders entail organizational policies and management practices.

---

[6]While procurement of PPT is one of the issues addressed in this study, the emphasis of this study is on obtaining the information needed to select the appropriate PPT rather than on funding problems or opportunities.

We use the term *community* to refer to the professional emergency responder community as defined by the types of organizations included in the discussions. The term *emergency responders* refers to those personnel within this community that deploy to emergency incidents. The term *first responders* was often used by participants in the RAND discussions; we use this term wherever it is valuable for highlighting issues that are salient to individuals who are the first to arrive at an incident scene.

SCOPE OF STUDY

Emergency responder organizations and specialties represented in this study include firefighting, law enforcement, emergency medical services, hazardous materials response, urban search and rescue (USAR), anti-terrorism, special weapons and tactics (SWAT), bomb squads, and emergency management. Note that this study did not include several actors that often serve in an emergency response capacity during particularly large events or when specialized expertise is required. Those actors may include municipal agencies and private organizations responsible for transportation, communications, medical services, public health, disaster assistance, public works and engineering, construction, and wildlands firefighting, as well as military elements such as the National Guard and the Army Corps of Engineers. As illustrated most recently by the September 11, 2001, attacks, the roles of such responders can be central in some cases (Jackson et al., 2002). However, because of the particular challenges involved in defining the roles and needs of workers who do not normally engage in emergency response, and because of the challenges presented by the diversity of practices, capabilities, and missions among these groups, evaluating the hazards and protection needs faced by "contingency" emergency responders requires a separate, dedicated research effort.

This study focused on obtaining input from responders and organizations at the local (city and county) level, given our interest in obtaining community views "from the field." Federal emergency response organizations were contacted to help provide background information on personal protection policy and technology research and development.[7]

In recent years, and especially after September 11, 2001, a number of efforts have examined emergency responder needs in a weapons of mass destruction (WMD) scenario (see, e.g., Dower et al., 2000; InterAgency Board for Equipment Standardization and InterOperability, 2001). This study endeavored to cover the

---

[7]Federal (i.e., Federal Emergency Management Agency) and state urban search-and-rescue task forces, who are major users of PPT, were not contacted as a group. However, because these forces are staffed largely by local firefighters and other specialists, the views of several USAR task force members regarding their USAR activities were noted during the discussions.

entire spectrum of operations undertaken and environments encountered—both usual and unusual—by local emergency responder organizations. In light of the September 11 terrorist attacks and the ensuing heightened attention to homeland security during the period when the discussions were conducted, the subject of terrorism preparedness and response was a prominent theme in many of those discussions. Nevertheless, participants' emphasized that needs for ongoing "conventional" operations must be considered along with needs emerging from unconventional operations such as for weapons of mass destruction scenarios.

ABOUT THIS REPORT

This report presents the results of RAND's discussions with 190 members of the emergency responder community concerning the risks they face in the line of duty and recommendations they made for enhancing their personal protection capabilities. The report conveys the views, experiences, and recommendations of the discussion participants. The emergency responder community is very diverse, and the discussions reflected that diversity. Accordingly, we have attempted to identify areas of consensus and disagreement and bring to light the implications of these perspectives for policymaking. We also highlight technology standards and initiatives from government and professional organizations that are germane to the issues and concerns raised in the discussions.

Before presenting the results of the community discussions, in Chapter Two we provide a brief overview of the emergency response community. This overview summarizes emergency responder organizational structures, emergency response activities, and injury and fatality data. The injury and fatality data complement the community views because the data can provide insights into the hazards that lead to injuries and deaths, while community views can help to identify those hazards for which the concerns are greatest within the emergency responder community. The hazards that emerge from the two sources are not always consistent. Protection from terrorism and protection from pathogens are two examples of concerns that are unimportant according to injury and fatality statistics but are nonetheless high priorities within the community.

Chapters Three through Six present the community views by major service lines:

- Firefighting (Chapter Three)

- Emergency medical response (Chapter Four)

- Law enforcement (Chapter Five)

- Hazardous materials and terrorism response (Chapter Six).

Within each of these chapters, we highlight the major risks, and the major health and safety and personal protection technology needs *at the individual-responder level* voiced by community members.

We encourage readers to review the findings in all of these chapters, not just those for their particular field of interest. While the conventional divisions among fire, EMS, and police are useful in terms of distinguishing professional career paths and primary job functions, there is considerable overlap among the services in the activities they perform, the hazards they encounter, and the education and training they need. Many personal protection issues—such as respiratory protection, personnel accountability, ballistic protection, and reducing the risk of exposure to pathogens, which were once relevant primarily to a single service—are becoming germane to all responders.

Expanding upon this point, many issues were raised in the discussions that do not fit neatly into conventional functional or service frameworks. These cross-cutting issues are presented in Chapters Seven and Eight.

Chapter Seven addresses *systems-level protection issues* such as communications systems and hazard information, identification, and assessment. Community members also illustrated how personal discretion and decision-making can be critical determinants of PPT effectiveness. To this point, the chapter addresses safety practices and enforcement, knowledge management (i.e., effectively utilizing available information), and the influence of service traditions and organizational culture.

Another crosscutting theme that emerged in the community discussions concerned the *organization and management* of personal protection: how PPT is selected, purchased, maintained, deployed, and retired. The findings in Chapter Eight highlight the centrality of PPT procurement and logistics in improving the personal protection of emergency responders when they are in the line of duty.

Finally, Chapter Nine extends the views and recommendations made in the discussions and presents several broad themes that may inform a personal protection agenda for the future for the entire emergency response community. The diverse and often complex issues raised by the community reveal a number of challenges for improving the personal protection of America's emergency responders, not just in the area of personal protection equipment, but also in the areas of risk assessment, education and training, information management and communications, and organizational development. Some issues can be addressed immediately through policy and program improvements, while others will first require analysis, research, and development.

Many of the points raised by emergency responders mirror issues raised at a NIOSH/RAND workshop that brought together personnel involved in the responses to the terrorist attacks of September 11, 2001, the anthrax attacks later that year, and the attack on the Murrah Federal Building in Oklahoma City six years earlier (Jackson et al., 2002). The information in this report and in Jackson et al., in addition to information on occupational injuries and deaths, will be used to develop a research and development road map for NPPTL. Further, given the wide range of organizations with a stake in improving emergency responder safety and health, we expect that a variety of other agencies and organizations will benefit from the findings of this study and act upon the ideas and challenges presented in this report, thereby better serving and supporting America's emergency responder community.

Chapter Two

OVERVIEW OF THE EMERGENCY RESPONDER COMMUNITY

The inherent risks and dangers in emergency response set it apart from most other professions. Compared with the average worker, emergency responders are about three times as likely to be injured or killed on their jobs (Clarke and Zak, 1999; Bureau of Labor Statistics, 2002). Compared with protecting workers in hazardous industrial environments, protecting emergency responders is particularly challenging because their working environment is varied and unpredictable, making it more difficult to catalog the risks they face and implement protections for them. The hazards that emergency responders face range from the mundane to the life-threatening and can change suddenly and considerably from day to day, incident to incident, and moment to moment.

This chapter presents an overview of the emergency responder community in terms of its size, activities, hazards, and injuries. This overview was compiled from data gathered during a comprehensive survey of publicly available sources and provides background for interpreting the community views presented in the subsequent chapters. It also provides an opportunity to examine the extent to which the views on hazards and protection needs expressed in the discussions compare with the available data on responder activities and injuries.

SERVICES IN THE EMERGENCY RESPONDER COMMUNITY

The emergency responder community examined in this study is typically divided into three services: fire, emergency medical, and law enforcement. While these three divisions serve as a useful classification scheme for discussing emergency responder career patterns and overall job functions, in terms of the community's activities and hazards, the boundaries among services are often blurred, especially the boundaries between fire and emergency medical services. This overlapping of activities and hazards is further complicated by the fact that a common mode of EMS delivery is through fire departments. Therefore, it is fairly common for firefighters to be cross-trained as emergency medical responders. A small fraction of law enforcement departments are also responsible for fire and emergency medical services, leading to additional overlapping. These overlaps in personnel and job functions led to some ambiguity in our compiling and interpreting statistics on the emergency responder community.

The Fire Service

In 2000, the United States had approximately 1.1 million firefighters working in more than 26,000 fire departments. About one-quarter of these firefighters were career (paid) personnel and three-quarters were active volunteers (Karter, 2001). These figures apply to municipal fire departments and exclude state and federal government agencies (which employ many wildland firefighters) and private fire brigades that protect industrial facilities.

Despite the fact that volunteers far outnumber career firefighters, 62 percent of the country's population is served by the latter (Karter, 2001).[1] While there has been a slow shift among firefighters from volunteer to paid status over the past decade and a half, the total number of municipal firefighters has remained nearly constant (Karter, 2001). Figure 2.1 shows the number and size of fire departments and the total number of firefighters as a function of the size of the population served.

While fire departments in the largest cities employ thousands of firefighters, most other departments are much smaller: More than 80 percent of departments protect populations of less than 10,000 and have an average size of fewer than 50 firefighters. As discussed later in this report, the decentralized structure of the fire and other emergency responder services makes it difficult for the emergency responder community to drive research and development which, in turn, impedes innovation and the flow of new technologies into the community.

---

[1]This results from the fact that populations protected by volunteer departments tend to have a higher ratio of firefighters to residents than those protected by career firefighter departments. This occurs primarily for two reasons: (1) compared with career firefighters, many more volunteer firefighters work part time, which requires a greater number of firefighters for a given population, and (2) most volunteer firefighters belong to smaller suburban and rural departments; because there is a minimum required size for a functional department regardless of the size of the protected population, volunteer departments protecting small populations have more firefighters per resident.

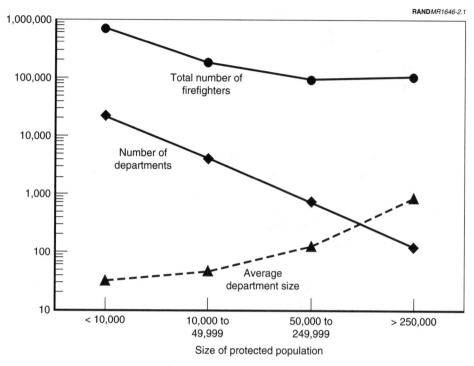

SOURCE: Karter (2001).

**Figure 2.1—Number and Average Size of Fire Departments and Number of
Firefighters in 2000**

The Emergency Medical Service

Because of the multiplicity of emergency medical service delivery systems,
emergency medical responders are difficult to count. One or more of a number
of organizations may provide EMS in a community. Those organizations include fire departments, independent third-service municipal agencies, hospitals, private firms, and law enforcement agencies. As a result, estimates of the
emergency medical responder population vary considerably. Our evaluation of
these estimates suggests that the population of active EMS responders may be
around 500,000.[2] Note that because of the common practice of cross-training

---

[2]Bureau of Labor Statistics (2003a) data indicate that there were 171,000 paid emergency medical
technicians (EMTs) and paramedics in 2001. However, the emergency medical services population
contains both paid and volunteer personnel. Data from the National Public Safety Information
Bureau (2002) and conversations with Bureau staff show that 5,885 EMS departments employed
about 212,000 emergency medical service responders, and 28,579 fire departments employed about
465,000 "emergency personnel," which includes emergency medical responders. Estimates of the

between fire and emergency medical service, there may be considerable overlap between the personnel included in this estimate and those in firefighting. As discussed in the following chapters, participants felt that the overlap of personnel and the multiplicity of agency types contributed to a lack of attention being paid to personal protection for the emergency medical response community.

Law Enforcement

There were nearly 800,000 full-time, sworn law enforcement officers in the nation in 2000, the most recent year for which such data are available for all levels of government. More than half of those officers were in local police departments, with the remainder in county, state, and federal agencies (see Table 2.1). Approximately 73 percent, or 580,000, of these officers can be considered emergency responders based on their primary responsibility for patrol duty or crime investigation.

In contrast to the fire service, the number of police and other law enforcement officers has increased steadily over the past decade. From 1990 to 2000, local

Table 2.1

Law Enforcement Agencies and Officers in the United States, 2000

Type of Agency	Number of Agencies	Number of Full-Time Sworn Personnel
Local police	12,666	440,920
County sheriffs	3,070	164,711
Primary state police	49	56,348
Federal	N/A	88,496
Special police and Texas constables	1,999	46,043
Total	17,784	796,518

NOTES: Special police agencies serve a special geographic jurisdiction or have special enforcement responsibilities; examples are campus, transportation, and parks and recreation police at both the state and local level. The number of federal agencies was not available; therefore, those agencies are not included in the total.

SOURCES: Federal number from Reaves and Hart (2001); all others from Reaves and Hickman (2002).

---

number of EMS certifications are relatively consistent at approximately 830,000 to 880,000 ("State and Province Survey," 2001; Heightman, 2000; National Association of Emergency Medical Technicians, 2002). Counts of certifications significantly overestimate the actual number of active responders because they include some emergency room and dispatch personnel and because certified individuals may hold more than one certification or not be working as EMS responders (Maguire et al., 2002). On the other hand, estimates of paid responders do not account for active volunteers. Taken together, these estimates suggest that 500,000 may be a reasonable estimate for the number of active EMS responders.

police, county sheriff, and state police officer levels increased by 21 percent, 16 percent, and 8 percent, respectively (Hickman and Reaves, 2001; Reaves, 1992). Federal law enforcement officers with firearms authorization and arrest powers saw an even larger increase—28 percent—from 1993 to 2000 (Reaves and Hart, 2001; Reaves, 1994).

As is the case with the fire service, the majority of police departments are relatively small: Nearly half of all local police departments had fewer than ten full-time officers in 2000. The size range of police departments is greater than that for fire departments, with police departments in the largest cities averaging more than twice the size of the corresponding fire departments, and police departments serving populations of fewer than 10,000 being four times smaller than the corresponding fire departments (see Figure 2.2). In contrast to the fire service, in which the vast majority of firefighters reside in small volunteer departments, police officers are more evenly distributed throughout departments of all sizes, with more in the largest departments (see Figure 2.2).

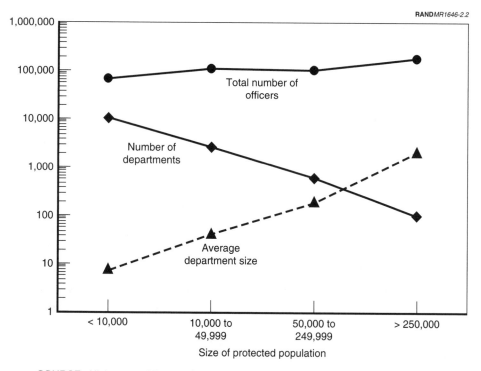

SOURCE: Hickman and Reaves (2003).

Figure 2.2—Number and Average Size of Local Police Departments and Number of Officers in 2000

Summary

The large number of small emergency response agencies in the United States, and the large variation in jurisdiction size and type, have great significance for the demand and supply of personal protective technology in the emergency responder community. Smaller organizations typically have greater funding constraints and more difficulty staying current on PPT information, acquisitions, and training. As discussed in subsequent chapters, RAND's discussions with the responder community revealed that the majority of personal protection innovation enters the fire service through the largest departments, and that smaller departments often turn to the larger services to stay abreast of information and trends. In addition, organizational heterogeneity combined with the highly dispersed and decentralized organization of firefighting, EMS, and law enforcement creates structural impediments to coordination within each sector, for instance, in the areas of PPT assessment, acquisitions, and deployment. Finally, organizational heterogeneity, combined with great uncertainty about potential risks in the future, results in widely differing views as to what are the most pressing PPT needs. Not surprisingly, a wide variety of perspectives were expressed in the discussions.

EMERGENCY RESPONSE ACTIVITIES

Fire and Emergency Medical Service

Over the past decade and a half, the role of the fire service has changed because of increased numbers of responses for emergency medical services, hazardous materials incidents, natural disasters, and terrorist attacks, and because of a substantial drop in the number of fire responses. In 2000, fire services responded to more than 20 million emergency calls; of those, about 60 percent were calls for emergency medical services and less than 10 percent were fire incidents. Between 1986 and 2000, the number of medical responses increased by 90 percent (see Figure 2.3).

In conjunction with their evolving role, fire departments are increasingly fielding specialized emergency response capabilities. Emergency medical service is the most common fire department specialization, with more than half of all departments and more than 75 percent of departments serving populations of 25,000 or more providing EMS response. Approximately 15–20 percent of fire departments maintain heavy rescue, hazardous materials response, and water rescue capabilities (Karter, 2001; National Public Safety Information Bureau, 2002). These specializations involve unique tasks, unique hazards, and special-

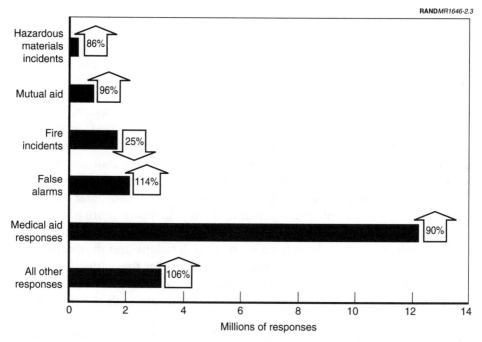

RAND*MR1646-2.3*

SOURCE: National Fire Protection Association (2002a).

Figure 2.3—Number of Fire Department Responses in 2000 and Percent Increase in Responses from 1986 to 2000

ized equipment, and therefore present new protection concerns. One of the major issues to emerge from the RAND discussions is the increase in specialized tasks that emergency responder operations must undertake and debates over the extent to which more-specialized personal protection equipment should be developed to better address the corresponding changes in hazard exposure.

Law Enforcement

Because police departments regularly patrol their jurisdictions as part of crime prevention and community interaction, a significant fraction of police activities is not initiated by a "call for service" in the same way that fire or EMS response is. As a result, a useful way to examine police activities is to consider all interactions between law enforcement officers and the public. A recent survey found that, of those citizens who had one or more contacts with police in 1999, 52 percent were involved in a traffic stop; 28 percent reported, witnessed, or were the victim of a crime; 21 percent asked for general assistance or reported a neighborhood problem; 13 percent were involved in or witnessed an accident; 3

percent were suspects in a crime; and 23 percent had other reasons for having police contact (Langan, 2001).[3] Although only about 1 percent of these interactions involved the use of force, the potentially serious consequences of the use of force make it an important consideration for officer safety and health protection.

Police departments may also perform more specialized functions, including bomb disposal, search and rescue, tactical operations (e.g., SWAT), underwater recovery, animal control, civil defense, harbor protection, and fire and emergency medical services. Most specialized capabilities are more common in larger departments: More than three-fourths of local police departments serving 250,000 or more people have bomb disposal responsibility, and more than three-fourths of local police departments serving 50,000 or more people have SWAT or tactical operations responsibility (Hickman and Reaves, 2001). Smaller departments are less likely to have such specializations and may depend on major jurisdictions nearby or state or federal assets for these functions. Some specializations, including fire and emergency medical services, are more common in smaller police departments. As is the case with the fire service, specialized capabilities involve a range of operations and hazards that differ from those in routine police work and have significant implications for protecting responder health and safety.

EMERGENCY RESPONDER INJURIES AND FATALITIES

Given the high levels of risk associated with their mission and the unpredictable aspects of their work, emergency responders face a broad range of hazards and are subject to significant numbers of occupational injuries, illnesses, and fatalities. The historical injury and fatality rates for police and career firefighters are approximately three times greater than the average for all professions, and place these careers in the top 15 occupations for risk of fatal occupational injury (Clarke and Zak, 1999; Bureau of Labor Statistics, 2002).[4]

Firefighters

Although firefighters undertake a variety of activities in the line of duty, those activities are not all equally hazardous. In particular, while firefighting calls rep-

---

[3]The percentages add to more than 100 because survey participants may have had multiple interactions with police or reported multiple reasons for an interaction.

[4]The relative risk to emergency responders while they are engaged in a response is actually considerably higher than this statistic indicates: Compared with conventional occupations that have a uniform level of risk throughout a work shift, the duration that emergency responders typically are engaged in a call for service or for an incident during a work shift is relatively short, thereby making that work interval extremely hazardous.

resent less than 10 percent of all fire department calls (National Fire Protection Association, 2002a), half of all firefighter injuries occur at fire scenes (Karter, 2000; National Fire Protection Association, 1995–2000). Of these fireground injuries, about half occur during fire attack, about 10 percent during ventilation and forcible entry, and about 16 percent during salvage and overhaul[5] (Karter, 2000).

Overall, approximately 88,000 firefighters were injured on the job each year from 1995 to 2000 (National Fire Protection Association, 1995–2000). Based on RAND's analysis of data gathered by the U.S. Fire Administration (1998), we estimated that more than 54,000 of these injuries were minor, while about 31,000 of these injuries were moderate, and 2,000 were severe or worse.[6] From 1990 to 2001, an average of 97 firefighters died in the line of duty each year (U.S. Fire Administration, 2002).[7]

The primary cause of both nonfatal injuries and death among firefighters is physical stress and overexertion (see Figure 2.4). Between 1995 and 2001, 45 percent of firefighter fatalities involved heart attacks (National Fire Protection Association, 1995-2001). These data are consistent with points that were made in the community discussions, which are discussed later, that physical and heat stress are critical hazards for firefighters.

Other major causes of nonfatal injuries occurring on the fireground include falls, exposure to fire products or chemicals, and being struck by or making contact with objects (see Figure 2.4). Of these other causes, exposure to fire products is the most serious, producing approximately twice as many severe fireground injuries as falling and being struck by or making contact with objects combined.

The other main causes of death among firefighters include becoming lost, caught, or trapped and motor vehicle accidents (see Figure 2.4). The risk of getting lost, caught, or trapped is another issue that figured prominently in the community discussions, with participants repeatedly emphasizing the need for improved fireground personnel accountability and personnel location tech-

---

[5]*Overhaul* begins when the main fire has been suppressed and entails activities such as searching for hidden hot spots, salvaging property, and cleaning up debris and equipment.

[6]The RAND estimate of injury severity is based on analysis of data from the National Fire Incident Reporting System, in which a *moderate injury* is defined as: "Little danger of death or permanent disability. Quick medical care is advisable." This category includes injuries such as fractures or lacerations requiring sutures. A *severe injury* is defined as a potentially life-threatening situation "if the condition remains uncontrolled. Immediate medical care is necessary." (U.S. Fire Administration, 1998).

[7]This average does not include responders killed in the terrorist attacks on September 11, 2001.

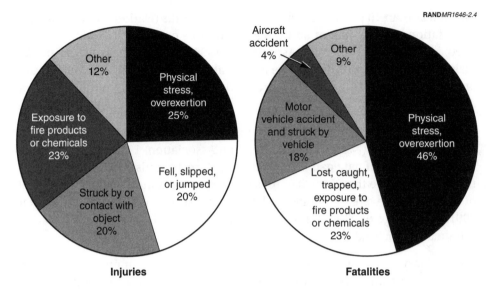

RAND*MR1646-2.4*

NOTE: Injury data are for fireground only.
SOURCES: Injury data are from an analysis of the National Fire Incident Reporting System Firefighter Casualty Module (U.S. Fire Administration, 1998). This database captures data for approximately 10 percent of all firefighter injuries. Only moderate, severe, and life-threatening injuries occurring on the fireground, as defined by the database, are included here. Assaults and vehicle accidents are included in the "struck by or contact with object" category, and "exposure to fire products and chemicals" is broken out from that category. Fatalities data are from National Fire Protection Association (1995–2001).

Figure 2.4—Causes of Firefighter Injuries and Fatalities

nologies. In sum, while being out on call represents only a portion of a firefighter's duty time, and fire calls account for less than 10 percent of calls for service, the fireground is an extremely high-risk zone.

Emergency Medical Responders

Because the labor force and activities of the emergency medical services are more difficult to define precisely than those of the fire service, injury and fatality data for emergency medical responders are more uncertain than the data for firefighters.

By far, the main cause of emergency medical responder line-of-duty deaths for which data are available is vehicle accidents. Our analysis of National EMS Memorial Service (2002) data indicates that there were at least 58 emergency medical responder line-of-duty deaths, or an average of about 11 per year, between 1998 and 2002.[8] We found that about half of all deaths resulted from rescue helicopter accidents, and approximately another third were due to ground transportation accidents or a responder being struck by a vehicle. An analysis of fatality data for 1992–1997 from three different databases by Maguire et al. (2002) gives a higher fatality rate: 114 deaths over 6 years, or 19 deaths per year.[9] Maguire et al. found a similarly high proportion of transportation-related causes: Nearly 60 percent were due to ground transportation accidents, and another 17 percent were caused by air ambulance crashes. Other major causes of fatalities were cardiovascular incidents (11 percent) and homicides (9 percent).

Among hospital-based emergency medical technicians, 18 percent of those whose records were publicly available reported exposure to potentially infectious bodily fluids between June 1995 and February 2002, with 1 percent being exposed more than once. About half of the exposures were due to percutaneous injuries, such as needle sticks, while the other half were due mostly to skin and mucous membrane exposures (Panlilio, 2002). Note that these numbers do not reflect actual infections.

In contrast to these data, which indicate that the primary hazards are vehicle accidents, heart attacks, and assaults, the primary concern among emergency medical service responders that was voiced during their discussions with RAND was exposure to infectious diseases. This discrepancy may reflect the status of current protective technologies: Decreasing the number of injuries from vehicle accidents and assaults may be viewed as being doable through better use of existing protective technologies and practices, whereas participants saw less possibility for greater personal protection from infectious diseases.

Law Enforcement

Assaults and physical stress each account for one-quarter of all police nonfatal injuries (see Figure 2.5). Other principal injury risk categories include falls (19 percent), motor vehicle accidents (16 percent), and being struck by or having contact with objects (10 percent).

From 1990 to 2001, an average of 155 police officers were killed in the line of duty each year (National Law Enforcement Officers Memorial Fund, 2002a).[10]

---

[8]Reporting of fatalities to the National EMS Memorial Service is voluntary, so the stated values represent lower bounds. This total and average do not include responders who are included in firefighter fatality data (presented in the previous section) or those killed in the terrorist attacks on September 11, 2001.

[9]Maguire et al. (2002) did not systematically correct for possible double reporting of fatalities in both firefighter and EMS data sets.

[10]This average does not include responders killed in the terrorist attacks on September 11, 2001.

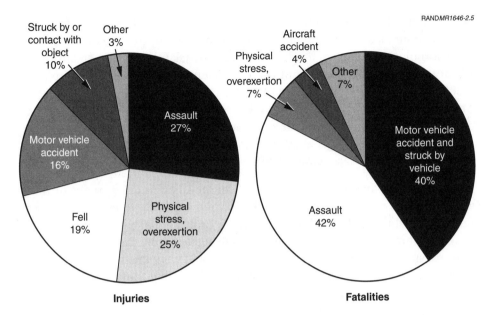

RANDMR1646-2.5

Injuries   **Fatalities**

NOTES: Injury data include data for police and detectives (occupation code 418), State of New York, at the local government level, for the years 1998–2000. Only injuries involving lost work days are included.

SOURCES: Injuries: Bureau of Labor Statistics (2003b). Some categories have been combined and/or renamed to match the definitions used by other data sources. Fatalities: National Law Enforcement Officers Memorial Fund (2002b).

Figure 2.5—Causes of Police Injuries and Fatalities

About 42 percent of line-of-duty fatalities are caused by assaults, and 44 percent involve vehicle-related accidents (including aircraft crashes and being struck by vehicles).

As discussed earlier, very few interactions between law enforcement officers and the public reportedly involve the use of force. However, the high incidence of police officer injury and death due to assaults suggests (albeit indirectly) that during the relatively brief periods of time they encounter hostile situations, law enforcement personnel are at very high risk of injury or death. This concern was expressed in the RAND discussions.

SUMMARY

The views expressed in the community discussions presented in the following chapters are generally consistent with the data cited in this chapter in terms of the activities that are most hazardous and the areas in which protection needs are greatest. However, because of the unpredictability of the risks faced by responders, it should be noted that injury and fatality surveillance data alone do not fully define responders' protection needs. Such surveillance data preferentially reflect the "routine" activities and hazards that occupy the majority of responders' time. The levels of injury are not, therefore, direct measures of the level of risk faced by responders for *all* activities.

Activities performed by responders for short periods of time or during events that occur infrequently could involve levels of risk much greater than the level of risk with more common tasks. Events such as a major disaster, structural collapse, civil disturbance, bomb disposal, hostage situation, or terrorism response involve hazards not normally encountered in routine activities. As a result, injury and fatality data do not effectively describe the potential consequences of such events, and the concern and attention they warrant for protecting emergency responders will not be reflected in those data. This concern was, however, apparent in the community discussions. Since September 11, 2001, attacks, the specter of terrorism involving the use of weapons of mass destruction has been central in guiding the priorities of the response community, particularly organizations in major urban centers. Such scenarios must be considered separately from more-routine responses, and preparedness needs for low-probability but very high-consequence incidents must be integrated into an overall protection strategy for the emergency response community.

Chapter Three

PROTECTING FIREFIGHTERS

When the RAND research team asked fire service representatives what primary occupational hazards and health and safety challenges their ranks faced in the line of duty, we received a wide range of responses. Despite the fact that calls for fires constitute less than 10 percent of service calls, they account for about half of firefighter injuries and fatalities (see Chapter Two), and the majority of hazards and protection needs identified in the discussions centered on the fireground. The principal areas of concern identified by fire service representatives included

- performance of turnout or bunker gear[1]

- heat stress while working in bunker gear

- respiratory protection and ways to improve the self-contained breathing apparatus (SCBA)[2]

- communications difficulties

- personnel command and control at the fireground

- logistical questions concerning personal protective technology management

- protection from chemical and biological hazards for front-line firefighters.

This chapter presents findings from the community discussions regarding many of these issues. (The community's concerns associated with medical calls are discussed in Chapter Four. Issues concerning PPT logistics and chemical/biological protection, which extend beyond the fire service, are discussed in

---

[1]Firefighter protective clothing, commonly referred to as *turnouts* or *bunker gear,* consists of flame- and water-retardant pants and overcoat.

[2]The *self-contained breathing apparatus* is a form of respiratory protection in which fresh air from a cylinder worn on the user's back is supplied via a pressure regulator and face mask.

Chapters Six through Eight, which address topics that cut across multiple areas of emergency responder protection.)

IMPROVING STRUCTURAL FIREFIGHTING ENSEMBLES

> Current thermal protection is adequate.
>
> We feel very well protected in what we are in.
>
> We are pretty well protected. Period.
>
> —*Fire service representatives*

One of the most consistent points raised in RAND's discussions with professionals engaged in structural firefighting is that their existing turnout garments provide excellent flame retardance and thermal protection—in other words, the protective capability of materials per se is not a significant concern in structural firefighting. However, participants pointed to some design shortcomings that diminish the overall performance of protective ensembles. Of these design weaknesses, the two that were most commonly raised were problems with component integration and compatibility and the poor functionality of gloves. Serious concerns were also directed at the need to reduce the heat and physical stress associated with working in modern bunker gear. Difficulties with inspection and assuring the integrity of protective equipment over time were other major concerns, and are addressed in this chapter and in Chapter Eight. Other concerns that were raised include ultraviolet degradation of thermal-protective components, difficulties with proper fitting of turnout gear, and inconsistent sizing of gear among different manufacturers.

Ensuring Component Integration and Compatibility

Our discussions with the firefighting community frequently focused on gaps in protection resulting from problems with the integration and compatibility of individual protective equipment components. These problems are manifested in three ways:

- Equipment performance degradation due to incompatibility of components

- Inconsistent protection levels among different components

- Bodily exposure at component interfaces.

One of the integration problems most frequently cited by participants is performance degradation and subsequent safety concerns resulting from incompatibilities of components. The most common examples cited were difficulties in performing certain activities while wearing firefighting gloves and being able to communicate while wearing an SCBA mask.

A second problem with component integration and compatibility cited by some participants concerns the inconsistent levels of fire protection among different components. For example, in the past, the hood was a weak link in the protective ensemble, so some firefighters used two hoods to compensate. Then, as hoods improved, the SCBA became the weak link. Now gloves apparently are, at least indirectly, the weak link.

A third integration-related problem has to do with the interfaces between components of the protective ensemble. One example of this cited by several departments is the use of gloves and turnout coats that have "wristlets" (elastic sleeves that extend beyond the end of the glove body or coat sleeve). Wristlets add protection by providing a tight seal and several inches of overlap between the glove and coat sleeve. However, due to incompatible designs, it can be difficult for the wearer to pull one wristlet over another. Therefore, when two wristlets are used, they can potentially bunch up such that neither is positioned properly. The result is a poor interface between glove and coat that may lead to skin exposure and the potential for burns or abrasions. This mismatch can occur because coats and gloves are often purchased from different manufacturers who do not necessarily integrate their designs. Other interface problems with turnout gear that were mentioned by participants include:

- Hood and SCBA: Mismatches may create gaps between the face opening in the hood and the SCBA mask

- SCBA and helmet: The SCBA tank can interfere with the rear brim of the helmet and can either restrict head motion or knock the helmet off; SCBA strap buckles can also interfere with proper helmet fit.

Both fire departments and equipment manufacturers mentioned problems with bodily exposures at component interfaces. Both groups noted, however, that many compatible interface options are available, and the problem can be addressed by careful selection of compatible components.[3] However, the potential still exists for selecting components with incompatible interfaces. Departments may make this error if they are unaware of incompatibilities between particular components or if they do not fully understand the functional

---

[3]It was observed by the community that the ongoing process of consolidation of PPT suppliers, whereby some manufacturers now produce entire protective ensembles, was leading to a reduction in the frequency of incompatibilities.

requirements of individual components in the context of the entire protective ensemble.

> It would be nice if manufacturers would standardize the connections outside of the [respirator] face piece.
>
> —*Fire service representative*

To address these issues, some departments advocated what one participant referred to as "configuration control." The goal of configuration control would be to reduce the potential for incompatible interfaces and modifications by requiring components to meet configuration standards (for instance, dimension, thread size, and fastener standards) in addition to performance standards. Configuration control would also make it easier for a department to upgrade equipment by ensuring compatibility over different generations of equipment. With configuration control, a department would be less likely to find itself in the position of not being able to upgrade to, for example, a new glove design because that design was incompatible with the turnout coats currently in service. Similarly, many participants expressed the desire for standardized configurations of radio component interfaces (e.g., batteries, harnesses, jacks), especially for different generations of devices made by the same manufacturer. Configuration control would also enhance equipment interoperability among agencies, which could have enormous value—both in terms of enhanced safety and cost savings in mutual aid[4] situations. (See Chapter Eight for further discussion of equipment interoperability.)

Concerns about component compatibility are beginning to be addressed through the introduction of standards. The National Fire Protection Association (NFPA) develops and maintains standards and certification procedures for a variety of firefighting and fire prevention equipment and protocols. For instance, the association maintains a standard for firefighter protective clothing, NFPA 1971 (National Fire Protection Association, 2000). Prior to 1997, this standard covered coats, pants, and hoods only. In an effort to establish a "systems approach" to the entire protective ensemble, this standard has been expanded to cover helmets, gloves, and footwear as well (which were previously covered under separate standards). While this standard still focuses primarily on the individual components and does not specify component interfaces, it is intended to lead to greater integration in design and more testing of firefighter protective clothing.

---

[4]A *mutual aid* response is one in which more than one department participates.

A related problem raised by some departments is the loss of equipment function as a result of adding accessories. To help differentiate their products, suppliers offer a range of turnout gear designs. Departments often request custom specifications on protective equipment for a number of different reasons—to address local working conditions, to promote organizational identity, and to adhere to tradition. In some cases, such customization may degrade the performance of the equipment. The following examples were cited:

- excessive knee padding that can limit the wearer's agility

- extra pockets, reflective trim, or lettering that can reduce heat loss and vapor transmission

- extra clasps and fasteners and excessively wide Velcro closures or storm flaps than can obstruct movement and have an increased risk of becoming entangled.

The firefighter protective clothing standard NFPA 1971 dictates that modifications influencing garment form or function require recertification to ensure that the component complies with the standard. Thus, customer-specified modifications should not result in equipment that fails to meet the standard. However, manufacturers are currently producing garments with materials that significantly exceed NFPA standards. Therefore, customization may degrade overall equipment performance while still meeting the standard. A problem associated with such customization is that purchasers are generally unaware of the degree of performance that is lost as a result of customization.

> A systems approach hasn't been applied very much to clothing. There has been a lot of dependence on the standard.
>
> —*Fire service representative*

One way to improve total ensemble performance, several participants noted, is by implementing full-body testing as part of a new certification process for garments, testing that would be akin to the current NIOSH respirator certification process. Current garment certification testing focuses primarily on the material properties of the individual components. Full-body/full-garment testing is complex and expensive, impediments that have thus far delayed its introduction. At a minimum, testing standards and protocols would need to be developed. However, one of the difficulties of full-body testing is the difficulty in being able to certify the innumerable permutations of separate components. Configuration control, the development of interoperability standards for uniform components, or the more extreme case of a standardized uniform would simplify the situation by limiting the number of ensemble component options that would require separate full-body testing. This discussion highlights the trade-off between ensuring ensemble performance on the one hand and promoting the innovations and alternative options provided by independent designs on the other.

Improving Gloves and Footwear

Gloves are a huge issue.

It seems like our gloves are ancient.

It would be nice to have a decent pair of gloves. Maybe [firefighters] would wear them more often.

If you can't do your work, you tend not to put them on.

—Fire service representatives

During their discussions with RAND researchers, participants raised many concerns about hand protection in the fire service. Many firefighters claimed that their gloves severely hindered or prevented them from performing important tasks such as manipulating radio switches, operating saws and other tools, or grasping SCBA mask straps. These problems stem from two principal concerns. The first concern was poor fit. Participants complained that glove fingers were too long, most notably the pinkie finger, which reduces dexterity, and that sizes and proportions varied greatly by manufacturer.

A second concern was glove materials. The thickness of the materials severely reduces dexterity. A technical expert familiar with glove construction and performance emphasized that currently available materials for gloves demand a trade-off between heat protection and dexterity. One consequence of this problem is the continuing use of non–NFPA-compliant gloves. For reasons having to do with habit, tradition, and improved dexterity, leather gloves are popular with many firefighters. But leather gloves do not provide levels of thermal protection commensurate with the rest of the firefighter ensemble. Additionally, firefighters mentioned that leather glove liners have a tendency to pull out or slip, especially when one is climbing a ladder. And when the leather dries, it becomes stiff.

Participants noted that because of severely degraded dexterity while wearing gloves, firefighters often must remove their gloves and sometimes suffer burns on their hands as a result. "I won't wear gloves for anything," said one fire service representative. Noting that most of the time firefighters do not require great dexterity, one participant advocated developing a two-layer glove system consisting of a medium-weight glove and a quick-to-remove-and-replace "overmitten." The mitten would provide maximum fire protection and sufficient dexterity for most operations and could be temporarily removed to allow for greater dexterity for certain tasks.

Since 1983, NFPA has published standards that cover gloves for structural firefighting. These standards address glove sizing and have resulted in improved fit and dexterity. The latest edition of these standards was published in 2000 (National Fire Protection Association, 2000). The 2000 NFPA glove standards include more-rigorous glove construction specifications and, for the first time, dexterity testing. Because this is a recent update, the firefighter community may not yet have had the opportunity to evaluate its effectiveness.

Footwear was also cited by many participants as an area needing significant improvement. For example, participants called for better traction and stability. Having consistent sizing of both foot length and width among different models and manufacturers was a need cited by one participant, who said, "It's amazing, you would think a size is a size, but it isn't so." Many fire service representatives reported the increased use of leather "ranger" boots: Several reasons were cited for wearing these boots, including better fit, greater comfort, and increased stability. The greater comfort of the leather boot was seen as an important benefit in lengthier responses.

Because standard-issue gloves and footwear are regarded as being so problematic, many fire departments reported that they allowed their personnel to buy and use their own gear under certain conditions.

Improving Gear Integrity and Maintainability

The performance characteristics of bunker gear degrade over time due to exposure to high temperatures, chemical exposure, repeated cleaning, exposure to light, and normal wear and tear. The participants in the firefighting community expressed concern with the lack of an objective means of determining when their gear required replacement. For example, many departments store their bunker gear in dark closets to minimize ultraviolet light exposure. But gear is exposed to sunlight on each fire call, and monitoring light exposure of equipment is not done. And even if light-exposure information were available, the rate at which exposure to light affects bunker gear performance is unclear. Participants raised similar concerns about repeated cleaning of gear, even though the cleaning conformed to the vendor's recommendations.

Exposing gear to very high temperatures results in an easily detectable loss of integrity; therefore, this is not a serious concern. What is more problematic is exposure to intermediate temperatures or to other impacts when damage cannot be visibly detected. Likewise, continuous use may result in compression be-tween layers in the equipment or damage that is not visually perceptible.

These concerns were validated during our discussions with technical experts familiar with firefighter bunker gear. They expressed their concern with the lack of test data on equipment integrity over time as a function of use, light expo-sure, and cleaning methods. Potential improvements in this area include ad-vanced materials that are resistant to photo degradation and imbedded sensors to detect equipment wear and/or failure. In an effort to address some of these concerns, NFPA recently introduced a standard for the periodic cleaning, main-tenance, and inspection of firefighter protective clothing (National Fire Protection Association, 2001a). In addition to promoting routine maintenance and inspection procedures, one of the objectives of this standard is the imple-mentation of a monitoring system to record cumulative exposures to various environments and cleaning cycles. Such records may provide valuable data for calibrating the impact of these various exposures on protective clothing per-formance. (For further discussion of maintenance and reliability of turnout gear and other protective equipment, see Chapter Eight.)

REDUCING PHYSICAL STRESS

Many firefighters with whom RAND spoke noted that physical stress is a serious hazard that arises, in part, as a consequence of the high level of performance of the modern protective ensemble. In addition to the weight of the SCBA pack, turnout garments are heavy, hot, and do not "breathe" well (i.e., do not allow body moisture to escape), which increases the stress on a firefighter. These problems may lead to, among other negative outcomes, heat stress, fatigue, scalding and steam burns, and shortened work cycles, especially during pro-longed events. The firefighters' concerns about physical stress are backed up by injury data. As was shown in Chapter Two, one-quarter of fireground injuries and nearly one-half of firefighter deaths are reported to be caused by physical stress.

The discussion that follows centers on hardware solutions, although it should be noted that several fire department representatives also stressed the impor-tance of procedural solutions, such as maintaining appropriate work cycles and improving training and adherence to conventional rehabilitation practices.

Improving Heat and Moisture Dissipation of Turnout Gear

> We want better heat stress relief.
>
> Anything you can do to make firefighters lighter and cooler . . . They are totally encapsulated and insulated.
>
> —*Fire service representatives*

One of the principal concerns voiced by firefighters in the RAND study was the risk of heat stress induced by increased body temperatures while fighting fires. The basic requirement that firefighters' clothing protect them from heat and flame necessitates designs that minimize heat transmission. This creates a trade-off between keeping firefighters cool and protecting them from burns. The NFPA protective clothing standard specifies minimum heat loss requirements, but most respondents agreed that improvements beyond the current standard were a high priority. One fire service leader noted that when his service adopted a more protective ensemble in the late 1980s, the number of burns decreased significantly, whereas incidents of heat stress increased. This challenge is especially acute in the Sunbelt states—the fastest-growing region of the country.

Several ways to address this trade-off problem were suggested to RAND researchers. Some departments have tried to alleviate heat stress by outfitting firefighters in shorts and T-shirts under their turnouts rather than traditional station uniforms. Some departments also reported trying different fabric materials for station gear. While these tactics appear to help, they can introduce new problems, the most significant of which is that a firefighter's arms and legs will be directly exposed to abrasions or cuts, body fluids, and other types of exposure during calls in which turnouts are not worn or are taken off, as is commonly done during emergency medical responses.

Another solution that was suggested is the use of active cooling technologies, such as ice packs and refrigeration systems. None of the departments participating in this study reported using active cooling for any tasks other than hazmat responses, although several expressed interest in the concept. The major hurdles to using cooling technologies that were cited include the added bulk and loss of mobility, the short useful life of ice packs (approximately 15 minutes), the burden of dealing with an additional piece of equipment, and cost.

A substantial portion of bodily heat loss occurs through sweat, and there was a strong desire for improved vapor transmission (breathability) of turnouts. Accordingly, the firefighting community put forth several potential directions for innovation, including the following:

- **Increasing the vapor transmission of turnout textiles.** Modern turnouts include a moisture barrier that repels liquid but allows sweat vapor to escape, facilitating heat dissipation. While significant progress in moisture barrier technology has been made in recent years, firefighters expressed a strong desire for continued improvement.

- **Improving turnout gear fit.** Many participants noted that turnout gear often is not well fitted to individual firefighters, even though manufacturers offer varied dimensions and fitting services. Participants called for improving the exactness of measurement services and for fully customized turnout construction.

- **Minimizing fabric layering.** Pockets, knee reinforcements, back supports, and other garment features all entail adding extra layers of fabric, which reduce breathability. And many fire services specify extra pocket space in response to demands for more storage areas. Improved garment engineering may be able to reduce fabric layering.

- **Reducing thermal protection in less-exposed areas.** Participants recommended reducing thermal protection in less-exposed areas, such as the armpit, the side of the torso, and along the back where the firefighter is protected by the SCBA frame. Textiles that provide variable levels of performance over different body areas are currently used in the construction of athletic gear.

Reducing Personal Protective Technology Weight

Firefighters repeatedly called for lighter, less-bulky garments. The weight of turnouts, in conjunction with that of other personal gear (SCBA, helmet, radio), amounts to a substantial weight burden on the firefighter. One participant pointed out that a fully equipped firefighter might be carrying personal protective equipment in addition to hoses, ladders, or power tools that total more than 50 percent of his or her body weight. Many participants felt that the bulkiness of turnout gear restricts their range of motion and therefore interferes with their firefighting ability. They noted that tasks such as climbing, crawling, and swinging tools were hindered.

Weight savings, it was argued, could be accomplished through several approaches, including

- more-customized fitting of turnout gear to eliminate unnecessary bulk

- integrating more functions into fewer pieces of equipment, such as an SCBA frame with an integrated antifall harness

- equipment design changes that minimize mass and weight
- use of lighter-weight materials.

The weight of firefighting helmets was a specific concern and is a good illustration of the potential for improvements from design changes and lighter-weight materials. One department noted that helmets were a leading source of chronic neck strain, and that it was becoming increasingly common for doctors to prescribe lighter helmets. Traditional leather helmets favored by many firefighters are heavier than more modern designs made with synthetic materials. Some participants raised questions about the appropriateness of the existing stringent impact-resistance standards for all helmets.[5] Several departments felt that the high-energy impacts that helmets are required to sustain under current certification standards would result in catastrophic injury to the wearer's neck and spine. Relaxing these standards, they claimed, could allow lighter helmets to be certified for service with little or no decrease in safety.

Does Encapsulation Increase the Risk of Injury?

I think we are going to a protective environment that is far beyond what the firefighter can deal with.

—Fire service leader

Another debate that surfaced in the RAND discussions was over the costs and benefits of encapsulation. Over time, personal protection for structural firefighters has evolved toward a state in which the firefighter is nearly completely encapsulated by highly insulating turnout coats and pants, gloves, boots, and protective hoods worn over the head and neck. While providing maximum burn protection, some participants (particularly senior members of the community) noted that this high level of protection has enabled firefighters to go deeper into a burning building, thereby increasing their likelihood of getting trapped. Ambient temperature, typically sensed on the ears, is one indicator used by firefighters to gauge hazards such as the proximity of a fire or potential flashover[6] conditions. In fact, participants noted that it was not uncommon for firefighters to suffer burns on an ear as a result of exposing the ear to gauge ambient temperature. Encapsulation, some fire service representatives claimed,

---

[5]The standard specifies that a helmet should withstand a test in which an eight-pound anvil is dropped from a height of five feet, imparting a maximum force of 850 pounds (National Fire Protection Association, 2000).

[6]*Flashover* is the near-instantaneous ignition of an entire room and its contents. Flashover is usually caused by the buildup of heat from a fire burning in one part of a room, which gradually heats the rest of the room to its ignition temperature.

has limited firefighters' ability to sense hazardous conditions. "We are sending our firefighters in too far," concluded one fire department leader.

Some participants countered that this situation was only the latest manifestation of continuing trade-offs between the intended protection and the unintentional drawbacks that emerge from protection innovations. Such trade-offs could be resolved, it was said, through the development and implementation of appropriate operational doctrine, training, and standards enforcement mechanisms. Others pointed to the development of turnouts with integrated temperature sensors and alarms as a potentially useful innovation. Most agreed, however, that existing technology solutions of this sort fall short in that they can be unreliable, for instance, should the sensor come into contact with hot surfaces. Interestingly, another concern is that the use of such technology would require the firefighting community to address several new questions, such as what the threshold temperature should be, what the minimum duration at that temperature should be, and what action a firefighter should take when an alarm is triggered.

Along these same lines, some participants went further to suggest adopting systems to perform *in situ* physiological monitoring of a firefighter's condition (e.g., temperature, pulse, respiration rate). Such systems would give firefighters a real-time, objective indication of their physiological status. If these data were then transmitted remotely via a radio system, safety officers or incident commanders could monitor their personnel and recognize when they were approaching the point of overexertion.

IMPROVING RESPIRATORY PROTECTION

Firefighters are exposed to a wide variety of gases, particulates, and other respiratory hazards in their activities. Respiratory protection, therefore, was a focus of attention during RAND's discussions with representatives from the firefighting community.

The SCBA is the standard respiratory protection technology used in the fire service. We did not hear of any use of air-purifying respirators (APRs)[7] in the fire service. While participants generally spoke very highly of the performance of SCBAs currently on the market, they did note some ways in which the technology could be improved to better meet their needs. A separate concern was that under current U.S. structural firefighting standards, the SCBA is the only means

---

[7]An *air-purifying respirator* comprises a half- or full-face mask with chemical-cartridge air filters. Ambient air is drawn through the filters, in some cases with the assistance of a fan (in powered air-purifying respirators), and is supplied to the user.

of respiratory protection that firefighters have available to them, but they questioned its appropriateness for all situations.

Improving SCBA Air Supply and Monitoring

A common desire expressed by participants was for a continuation of the trend toward lighter, more-compact SCBA tanks with higher capacities and longer air supplies. Many agencies reported acquiring latest-generation bottles with extended air supplies. This is of particularly great importance in extended emergency response campaigns, such as after the September 11 Pentagon and World Trade Center attacks, where SCBA air supplies lasted far shorter than the duration of the response (Jackson et al., 2002).

Participants also called for better air supply monitoring and low-volume warning systems to get more effective use of the air supplies they have available. Some respondents stressed the importance of providing continuous monitoring of remaining breathing time rather than a sudden warning at low tank pressure. Continuous monitoring allows responders to plan their actions accordingly and avoid getting caught in a situation in which it may be inconvenient or dangerous to exit an area immediately. A low-pressure alarm is also inexact because the relationship between pressure and remaining breathing time depends on the user's rate of respiration, which varies from person to person as well as by the level of activity in which the user is engaged. Thus, the remaining air supply duration for a firefighter engaged in a very strenuous activity may be substantially shorter than that for a firefighter engaged in a less physically demanding one. A better approach, participants claimed, would also monitor the rate of pressure change to more accurately calculate remaining breathing time. Participants also expressed the importance of a visual low-air-supply alarm, because audible alarms are sometimes difficult to hear amid the background noise at a fire scene.

To these points, many participants applauded the recent improvements in the NFPA SCBA standard, which have begun to address these concerns. The new standard (National Fire Protection Association, 2002b) requires in-mask visual display systems that indicate remaining air pressure in multiple increments of total rated tank pressure, providing firefighters with a gradual warning of their remaining air supply rather than just a single low-air alarm.

Expanding Options for Respiratory Protection

Many departments noted that while SCBAs provide a very high level of protection against airborne hazards, they are not always appropriate for the varied tasks and levels of risk firefighters face in the line of duty. Some participants stated that there are situations in which they would like to have alternative respiratory protection options.

The most commonly cited situation in which alternative options would be useful was the overhaul stage of fire suppression. Participants indicated that while department policies typically mandated that SCBAs be used during overhaul, compliance is low. This lack of compliance stems from the view that an SCBA is often overkill during this stage and from problems with ensuring that enough air remains at the end of an event. Accordingly, many participants expressed a desire to use an APR with the appropriate filtration capability. Because an APR presents less physical inconvenience than an SCBA, firefighters would be more inclined to use it. One variant that several departments mentioned in a favorable light is a currently available dual-purpose mask in which the SCBA regulator and hose can be disconnected and replaced by an air-purifying cartridge. Several participants questioned this recommendation. They noted that while a convertible mask allows the user to shed the SCBA tank, it does nothing to alleviate the restricted vision and comfort problems associated with the mask. Manufacturers noted that there are also some technical difficulties in designing a mask that can be used for both positive pressure (SCBA) and negative pressure (APR) applications. One department reported having conducted a study in this area and found that the currently available air-purifying respirators do not provide adequate protection from the hazards encountered during overhaul, such as toxic gasses emitted by building interiors and furniture. "Our greatest exposure occurs *after* the fire," said a department representative.

Thus, while the present strategy of SCBA use during overhaul suffers from a lack of compliance, the appropriate solution for maximizing their health and safety was not obvious to participants. (For a further discussion on protective equipment tailored to specific hazards, see Chapter Eight.)

Another situation in which APRs were viewed as a superior respiratory protection option is during search-and-rescue campaigns after structural collapses. Findings from emergency response to the World Trade Center attacks show that SCBAs are too cumbersome and their air supplies are too limited for such operations. APRs, on the other hand, were widely distributed at the attack site and were viewed as an essential protective measure (Jackson et al., 2002).

IMPROVING COMMUNICATIONS CAPABILITIES

Participants repeatedly pointed out that firefighters have great difficulty communicating while wearing an SCBA face mask. Firefighters noted that sound travels poorly in smoky air, and that there is substantial background noise at fire scenes (from engines, sirens, alarms, pumps, hoses, power tools, and other sources). Department after department relayed how firefighters typically have to yell and repeat themselves to be heard by a person standing right nearby. Efforts to mitigate this problem through retrofitting masks with mechanical or battery-powered voice amplifiers were seen as providing only marginal improvements. With these systems, the user's voice is reflected inside the mask, such that a firefighter's speech still sounds muffled, only louder, observed one participant. When communicating with each other, firefighters often have to remove their face masks or resort to hand signals.

Moreover, poor radio communication between firefighters and incident commanders is also considered to be a serious problem by most of the community. Technical advances to address this problem include improved radio microphones, the positioning of these microphones directly on the side of the face mask, and specialized earphones, although departments reportedly are not entirely satisfied with these solutions—both in terms of improved communications and functionality. As one participant put it, they are "not as good as you'd think. They involve extra wires, extra cost, and communication is not that clear." Departments reported difficulties with the microphone cables getting snagged on debris, tools, and various other items. One department abandoned the use of integrated respirator and communications systems in part because firefighters grew tired of having to connect the cables each time they donned their gear. Two solutions that participants cited were using a wireless link from the mask to the radio unit, which would be mounted on the body or SCBA frame, or setting up a wireless local area network to support sitewide communications and data transfer.

Numerous fire department representatives also expressed dissatisfaction with the radio units themselves. The representatives voiced a wide range of complaints, including

- small controls that require personnel to take off their gloves, increasing their risk of getting burned

- controls that are susceptible to being inadvertently switched when the unit is bumped or swiped

- incompatibility of jacks and other components among models made by the same manufacturer and different manufacturers

- insufficient battery life

- inadequate water and thermal resistance.

Many of these of these problems, the participants noted, could be remedied relatively easily without substantial research and development costs. Several participants expressed the view that these problems persisted because radios used in the fire service are not necessarily designed with the firefighting mission uppermost in mind. A communications solution optimized for firefighters may not be easy to obtain, they claimed, because firefighting represents a small share of the market for radios.

Another communications issue that was raised is the desire for a dedicated transceiver for responder emergencies, such as mayday and evacuation calls. Participants were concerned that mayday calls can get lost in radio "chatter." The existing system for mayday calls includes a button on the radio unit that opens a channel to the dispatch center and identifies the unit to which the sending radio was assigned. The dispatch center then contacts the appropriate scene commander based on unit assignments. The evacuation call system is essentially the reverse of the mayday call process: An incident commander issues an order, which is then relayed by the dispatch center to the individual firefighter. An improved system would comprise a dedicated device that would enable a responder who is in trouble to send mayday calls and receive evacuation calls directly to and from incident or unit commanders. Advocates for such a device stressed that for this technology to have maximum effectiveness, it would need to be physically separate from the voice radio, would always have to be on, and would use reserved radio frequencies and distinct warning tones.

Finally, the issue of the cost of improved communications was raised by many participants. Although, in law enforcement, each individual patrol officer typically is assigned a radio, this is not the case in firefighting or medical response: On average, fire departments have only enough portable radios to equip about half of the emergency responders on a shift (U.S. Fire Administration and National Fire Protection Association, 2002). Thus, for many fire departments, their priority is in increasing the number of responders having radios (necessitating a focus on lower unit costs) as opposed to increasing radio capabilities (likely resulting in higher unit cost).

IMPROVING PERSONNEL ACCOUNTABILITY

Another top-priority issue that participants raised about firefighter safety concerns on-scene personnel management. Many firefighters are injured or do not receive prompt treatment for injuries, participants claimed, because of confusion about the location and activities of individuals at an incident. Even when all responders are in radio communication contact, it is often difficult to know where individuals are relative to one another. This problem becomes more serious in situations in which injury or signal loss prevents communication. For these reasons, participants expressed a desire for the ability to monitor and manage the precise location of personnel, independent of reliance on voice communications. During the RAND discussions, participants cited several potentially promising solutions to such concerns. (Further discussion of personnel location tracking is in Chapter Seven.)

A less technically demanding way to track firefighters at an incident is with fireground personnel accountability systems. These systems are designed to keep track of whoever has been deployed to a scene and what tasks each individual is doing. Nearly every fire department raised fireground accountability as an area in need of improvement. Present systems typically involve personal identification tags and status boards. Individuals transfer these tags, which are typically attached to either their helmets, bunker coats, or radios, to one or more status boards as they arrive on a scene and engage in a particular task. Participants noted that this system, while simple, is susceptible to errors. Also, even when used properly, the system provides only limited information on an individual's location, which is inferred from the task they are assumed to be performing. A firefighter from one department described the accountability tags as being archaic and highly ineffective. In addition, many departments use accountability tags only for "special hazard" responses, such as confined-space rescues and attacks on fires in high-rise buildings, and not on a routine basis.

Several participants suggested improvements to the existing personnel accountability system, and noted that promising new approaches are under development and may soon become available. These new approaches include card-swipe or bar-code systems, in which portable magnetic card readers or bar-code readers are positioned at specified points around an incident scene. Also suggested were identification-tag reader systems, analogous to antitheft tags used in retail stores, which would passively register when a responder passes a specified point. An advantage of these systems is the potential for setting up as many registry stations as needed. This capability would allow commanders to define specialized task and location categories as needed for a particular incident and would also provide higher-resolution information on the location of individual responders. Another advantage to such a system is that it would update information automatically—for instance, when a responder registers at a second station, he would be unlisted from the first station—thereby minimizing accounting errors. These systems could also support remote data access, allowing commanders at major incidents to monitor and control operations a greater distance away.

Such technologies would greatly improve fireground safety, although there are serious barriers to their implementation. In addition to the investment costs and maintenance concerns, the systems must have the capability to be set up quickly and operate reliably in harsh environments. In addition, they should be designed to operate as "open" systems, allowing responders from multiple departments and services to utilize a single system at an incident.

Chapter Four

PROTECTING EMERGENCY MEDICAL SERVICE RESPONDERS

As shown in Chapter Two, the United States has seen a long-term rise in the number of emergency medical calls. Discussion participants expected this trend to continue as health care costs increase and the nation's population ages. Moreover, as with firefighting responses, participants observed that the complexity of emergency medical incidents is increasing. Given these trends, emergency medical service response personnel pointed out that they are experiencing a concurrent increase in the dangers they confront while lacking the appropriate personal protection to safeguard against those dangers.[1]

Among the concerns voiced by emergency medical service responders during their discussions with RAND, uppermost was limiting their exposure to infectious diseases. As with all emergency responder services, terrorism was also a major concern among EMS responders. Many participants, especially in the larger departments, also expressed concern over the increasing threat of assault. EMS personnel also noted that addressing the protection needs in their service is hindered by the multiple types of agencies engaged in EMS response. Participants felt that this heterogeneity in EMS agencies tends to reduce the visibility of the emergency medical service and limit the amount of guidance and support it receives from government and professional organizations.[2]

---

[1]The term *emergency medical responder* refers to both emergency medical technicians (EMTs) and paramedics.

[2]The findings for EMS responders must be qualified by noting some sampling limitations. Nearly all of the fire departments RAND contacted are the primary EMS providers for their jurisdictions, and representatives specializing in EMS were present in most fire department discussions. Three independent emergency medical service providers were also included. However, nationwide, only about 40 percent of EMS response is provided by fire departments (Karter, 2001), with independent agencies and, to a smaller extent, hospitals, private firms, and law enforcement agencies making up the remaining 60 percent. The input in this study is thus biased toward the fire service. The potential implications of this bias for the findings are unclear. The integration with firefighters and associated emphasis on safety and access to quality equipment may lead to fire-based EMS systems experiencing fewer shortcomings in their PPT options.

LACK OF SPECIALIZED PERSONAL PROTECTION TECHNOLOGY FOR EMERGENCY MEDICAL SERVICE RESPONDERS

According to the personnel with whom we spoke, few emergency medical service teams have an adequate supply of personal protective technology on hand and ready for use. Like law enforcement personnel, emergency medical responders often are the first on the scene of an emergency and, therefore, must use whatever PPT is on their vehicles. "Medical responders don't have anything," one participant pointedly said. PPT training reportedly is also in short supply among EMS personnel. For example, one agency provides its personnel with only 12 hours of PPT training. "How we are going to follow up on this, I don't know," said a representative of that agency. "All of our training is done on overtime. It's an expensive proposition. We gave up a lot of other things [for training]."

To remedy the situation, some organizations are adopting PPT, such as SCBAs, bunker gear, armored vests, and practice standards, from the fire service and from law enforcement. Two independent (third-service) emergency medical response organizations reported issuing all medical response personnel standard fire-rated bunker gear. One reason they cited for doing this was that EMS personnel often are trained for and serve in the fire service and therefore were already issued gear. Also, emergency medical service personnel, regardless of their organizational affiliation, often participate in technical rescues[3] and play an integral role at fire scenes. Finally, the cost differential between fire garments and EMS garments was not seen as being substantial. "We went for the optimum protection factor," said a representative from one service. "We wanted to make sure we exceeded the NFPA standard for thermal protection." Biological contamination of bunker gear was not seen as being a problem, claimed one participant, because emergency medical responders "have a better understanding of blood-borne pathogens" than firefighters, and could therefore manage such contamination with proper cleaning and care of the gear.

> EMS has been very underequipped for a very long time . . . [PPT] has not been a high priority in our industry.
>
> —*Emergency medical service leader*

More fundamentally, protective gear used by emergency medical responders is often not developed specifically for their jobs and the hazards they face. One participant, for example, noted that the protective gear that is currently avail-

---

[3]*Technical rescue* is a class of emergency response that typically involves special-access capabilities, such as searching, climbing, repelling, or moving heavy objects.

able creates "plastic bag syndrome," meaning that it made responders feel like they were working inside a plastic bag. "We aren't going to wear this stuff on a regular basis," he claimed. Another representative noted that even though his agency was seeking to provide a high level of respiratory protection for its medical personnel, WMD exercise scenarios had indicated that "there is no way we are effective in SCBAs."

One reason cited for these protection shortcomings is that no federal agency is dedicated to addressing the hazards and protection needs of the emergency medical responder community, and little funding is dedicated to address these issues. Two participants cited efforts by the U.S. Department of Transportation in the 1960s as important first steps in improving equipment for emergency medical responders, but those efforts were seen as now being outdated.

A second reason for these shortcomings is that despite the fact that emergency medical service response in many medium-size and large jurisdictions is provided by fire departments, and that three out of four fire service responses are for medical aid, the National Fire Protection Association's substantial efforts and influence regarding protection and safe practices have been focused primarily on organizations and personnel whose primary training, operations, and institutional culture are focused on structural firefighting. For example, while the National Fire Protection Association (NFPA) maintains a standard for protective clothing for emergency medical operations (National Fire Protection Association, 1997), protective clothing meeting this standard reportedly is not in widespread use. The reason for this may be because, as some participants claimed, the clothing does not adequately meet the needs of emergency medical responders. In addition, participants claimed that the standards for emergency medical response operations (NFPA 1710 and 1720) and technical rescue operations (NFPA 1670), while providing a good basis for service provision, were not being implemented evenly, in part because the standards do not fully address the organizational needs, practices, and priorities of emergency medical services. To illustrate this point, a big-city fire department representative who spoke with RAND argued that these standards were too rigorous and too expensive to implement.

Complicating the problem of inconsistent application of standards for protective technologies are the diverse types of organizations that provide emergency medical service response and the inevitable variations in practices and procedures that they follow. In addition to local fire departments, many communities have independent municipal, private, or hospital-based emergency medical services. "There are so many flavors of EMS out there," observed one community member. As a result of this structural heterogeneity, service practices vary significantly within the emergency medical response community. Unlike firefighters, who are expected to arrive at a fire scene wearing bunker gear, medical personnel have much greater latitude in the personal protective equipment and practices they use, depending on local policy, the nature of the event, and individual discretion. For example, according to representatives with whom RAND spoke, the acquisition and use of ballistic vests often is left to the discretion of individual responders.

Finally, when compared with the fire service, emergency medical service responders operate more autonomously and typically do not have chiefs or other safety personnel on the scene enforcing PPT use. One department issued its medical responders fanny packs containing a particulate mask, goggles, gown, gloves, and scissors. But, he added, "It has taken a long time to get people to wear them on the majority of calls." Another fire-based service issued particulate masks to its personnel, but one representative of that service noted, "You'll see some crews religiously wearing them on their shoulders."

HAZARDS AND TECHNOLOGY PRIORITIES FOR EMERGENCY MEDICAL PERSONNEL

The wide variation in the organization of and management within the emergency medical services makes characterizing the service's practices and priority technology needs more difficult than characterizing those of other services. In this section, we outline several key issues related to health and safety risks and the technology needed to address those risks that were raised by emergency medical responder representatives.

Protecting Against Pathogens

Although a substantial fraction of emergency medical service responders are exposed to potentially infectious bodily fluids, surveillance data indicate that fluid-borne pathogens are not a major cause of injury or death among emergency medical service responders (see Chapter Two). Nonetheless, exposure to liquid-borne and airborne pathogens is the principal concern among emergency medical service responders, according to medical personnel that RAND contacted. Participants mentioned their concerns about increasing threats of exposure to hepatitis C, human immunodeficiency virus (HIV), tuberculosis, meningitis, West Nile virus, and childhood diseases. Some departments noted that the air inside ambulances can become particularly hazardous during transport of some patients. These concerns were not just confined to major urban services: Representatives from affluent, suburban, and rural communities also spoke of these issues.

Emergency medical response services have long had basic protective gear at their disposal, such as latex gloves, particulate filter masks, eye protection, and gowns. Most respondents maintained that the level of protection this equipment provides is adequate, when it is used. In line with the concerns mentioned in the previous section, such gear largely is designed for hospital-based care and is not specially developed for use in the field. For example, providing splash protection for the forearms of responders wearing short-sleeved shirts in warm weather conditions was seen as a particularly intractable problem. "I just don't know how you protect yourself," said one representative. Hands and forearms were seen as the most critical points of exposure, with exposures to broken skin, puncture wounds, and bites that penetrate gloves being common risks. The representative just quoted reported that his agency provides its personnel with a day of self-defense training to reduce the likelihood of exposure to risks and injury.

Addressing Increasing Concerns About Assaults

> Attacks on emergency responders are increasing.
>
> —*Emergency medical service responder*

Another concern voiced by EMS responders in their discussions with RAND was their increasing concern over being physically assaulted while on the job. Emergency medical responders noted that they must operate in numerous types of situations and under unpredictable circumstances. Unanticipated criminal activity, domestic violence, hostage situations, and abusive or mentally ill patients are just some of the ancillary hazards they may encounter at incident scenes. A representative of a third-service department said that protection from assaults was his greatest concern. To address this concern, personnel in his department were given self-defense and situation-management training.

Although emergency medical service responders in several larger departments reported that they have been issued ballistic vests and jackets, use of such gear is estimated to be rare. A representative from one large urban department estimated that responders wore armored garments on less than 1 percent of medical calls. In all cases, use of body armor was left to the discretion of the individual; when it is worn, respondents noted, it is often on nighttime calls only. One department noted that it used to provide ballistic vests to emergency responders, but could no longer do so for cost reasons. Standards had changed such that the vests had to be issued and fit to specific individuals, requiring the purchase of more vests than the department could afford.

Seeking Greater Protection from Weapons of Mass Destruction and Chemical Threats

Exposure to anthrax and other biological and chemical agents has also become a primary concern of medical responders in this post-9/11 period. In the words of one representative, "This is a whole new ball game." In the event of a chemical disaster response or terrorism response, emergency medical responders are expected to enter the affected area, tend to victims, assist with their extrication and decontamination, and manage their care until they are delivered to a medical facility. Emergency medical service personnel play an additional critical role in rendering medical assistance to injured emergency responders. With this in mind, responders raised concerns about their potential direct exposure to chemical and biological agents and secondary exposure to these agents through contact with contaminated individuals and materials. At the World Trade Center site in 2001, for example, responders repeatedly were coming into contact with body parts during the recovery stage. "Prior to 9/11," said one participant, "public health had never been a priority in these incidents."

In response to increased concern about such threats, emergency medical service responders reported that their organizations have sought to enhance personal protection capabilities. One EMS representative noted that while all personnel in his department had full bunker gear, none had access to SCBAs, a shortcoming that the department had recently recognized as one needing to be addressed. Another service opted to place two SCBAs on every truck in its fleet and issue responders face pieces that also accept air-purifying filters to be worn during extended response times in the case of a hazmat event or chemical attack. The service also equipped its trucks with chemical protective garments, duct tape, and chemical-resistant Nytril gloves. Another EMS service outfitted its vehicles with personal protective equipment kits to be used in the event of a WMD attack: large duffel bags containing gas masks, emergency escape hoods,[4] dust masks, helmets, goggles, and leather gloves. The service's goal was for each of its EMS squads to be able to handle 25 to 35 patients immediately in the case of a WMD attack, pending the arrival of backup support.

A critical part of taking the proper precautions in a WMD or chemical threat environment, two EMS agency representatives observed, is having an awareness of the potential threats in such environments. Yet, unlike the fire service and law enforcement, which have specialized teams such as hazmat and SWAT teams that possess special training and equipment to deal with nonconventional and extremely hazardous situations, emergency medical services for the

---

[4]An *emergency escape hood* is a soft-sided pullover hood with an elastic neck seal. These hoods provide particulate and chemical respiratory protection enabling wearers to exit hazardous environments.

most part have not developed their own hazard-awareness protocols, training, and capabilities, even though they are often among the first responders at incident scenes. One solution to this shortcoming cited by participants is providing emergency medical service personnel with environmental monitoring technologies, such as indicator badges that would alert them to hazardous conditions. A more immediate solution is interagency training exercises, through which EMS personnel would be able to quickly learn hazard awareness and response skills and develop protocols for sharing information and coordinating activities, such as for hazmat, with other services that already have advanced capabilities. The goal of such efforts, one participant said, was to avoid the "rush-in mentality" and stage emergency response operations from a safe place. "We stress that over and over again. We just want our people to live through it."

Chapter Five

PROTECTING LAW ENFORCEMENT RESPONDERS

One conclusion that emerged from our discussions with the emergency responder community is that protecting the health and safety of law enforcement responders may be the most challenging personal protection task within the community.[1] A major part of the challenge stems from the difficulties in characterizing the hazards that law enforcement responders face. These difficulties are compounded by the fact that law enforcement personnel are typically the first on the scene of an emergency or incident, and therefore have the least amount of advance information about the scene's potential hazards. In addition, the range of hazards that law enforcement responders face continues to increase, with exposure to infectious diseases and terrorism ranking as the most important concerns.

CHALLENGES OF PROTECTING LAW ENFORCEMENT RESPONDERS IN THE LINE OF DUTY

> In our little "burg" we have enough bizarre [situations] . . . . It's usually the case that we don't know what we are getting into until we get into it.
>
> —*Law enforcement representative*

Several factors affecting the use of personal protective technology in law enforcement were raised by law enforcement representatives in their discussions with RAND.

First, patrol officers often are the first to arrive on a scene and are expected to render assistance while maintaining law and order. "We are trained to drive right up to something and fix it," said one police officer. In many cases, patrol

---

[1]The focus in this chapter is on "main-line" law enforcement personnel (e.g., police, sheriffs, state police, transit police). Personal protection for specialty units such as SWAT, bomb, and anti-terrorism squads are addressed in Chapter Six.

officers discover unsuspected health and safety hazards only while in the process of being exposed to those hazards. An example cited by participants is the witnessing of a violent assault or the discovery of a methamphetamine laboratory when responding to a domestic disturbance call. "[Methamphetamine labs] are unlike anything we have dealt with before," said one official of a mid-size city, voicing a concern that was echoed in several discussions. In talking about this issue, many participants referred to police officers as "blue canaries." Said one commander, "It's funny, but it's probably true." But, he added, in the post-9/11 environment, "It's not stuff we can laugh at any more."

Second, given their need for agility, flexibility, and speed, police officers cannot be burdened with excessive or restrictive gear. Situations in which PPT could impair an officer's performance include foot pursuits, the use of firearms, and physical altercations. Recent changes in operational doctrines that emphasize a more-proactive, offensive response to threats in the community have put mobility at a premium, several law enforcement representatives observed.

> The handier you make it, the more likely you are to use it.
>
> Nobody has been able to design gear for the range of environments that police find themselves in.
>
> —*Law enforcement representatives*

Third, most law enforcement personnel are on patrol in the field between calls and therefore usually have very limited personal protection equipment that they can grab quickly in the event of an emergency. The trunk of a patrol car has proven to be inadequate for storing PPT (see Chapter Eight for further discussion of this topic). Patrol officers on foot or on bicycle cannot carry much gear with them at all. Yet, they encounter widely diverse environments and scenarios over the course of a work shift. Unlike the fire and medical services, law enforcement agencies typically do not have backup technology resources they can call on: "You can be reasonably assured that [firefighters] have the gear they need," or they can call in another truck that has the gear, said a police official from a mid-size city.

Fourth, responders' appearance is a concern, which places another constraint on developing PPT for law enforcement. Because of their frequent face-to-face contact with citizens and the increasing importance of fostering and maintaining strong ties with the communities they protect, patrol officers and other law enforcement personnel should not be burdened with excessive gear, especially gear that can be perceived as being threatening. Undercover agents and anti-terrorism squads need to blend into their surroundings and not become targets of attention.

We have not been able to provide adequate continuing education. Our continuing education and enforcement stinks.

—Law enforcement representative

Fifth, time for training is limited, which creates another impediment to the effective use of PPT in law enforcement. In contrast to the fire service, in which station time can be used for conducting training and refresher courses, patrol duties and case loads leave police officers little time to obtain extensive training in areas such as hazard identification, use of personal protection equipment, and safe practices. Although many agencies noted that several extramural training opportunities are available, particularly in the areas of terrorism and WMD response (much of this effort is coordinated through the State and Local Domestic Preparedness Training and Technical Assistance Program of the Office of Domestic Preparedness), many agencies lack the money for backup personnel to fill in for officers who are pulled off their shifts for training.

Sixth, most personal protective equipment and practices are not developed with the law enforcement mission and operating environment in mind. Compared with the fire service, law enforcement in the United States has fewer guidelines, standards, institutions, and committees addressing its protection needs. "We don't have a whole lot of regulations," said one police official. The National Institute of Justice (NIJ) maintains an active program of research on and standardization of technologies for law enforcement, but, aside from research on ballistic vests, little of this effort is directed toward personal protection. Where it is feasible, many agencies observe NIOSH and NFPA guidelines.[2] Yet, implementation of these guidelines might still be haphazard because most law enforcement agencies have not put into place the safety committees, compliance officers, and enforcement practices that are common in the fire service. "There's a basic lack of oversight on this job," said one law enforcement representative.

Cops have absolutely nothing to keep themselves safe.

—Law enforcement representative

Finally, PPT has not historically ranked as a critical policy, management, or budgeting priority in the law enforcement community. In talking about ongoing efforts to upgrade PPT for a WMD event, a police official in one mid-size city

---

[2]One participant noted that law enforcement personnel who also serve as volunteer firefighters and attend firefighting seminars can be important conduits of the latest information on PPT.

commented, "We are the only public servant first responder [organization] that has never been mandated to have such equipment." A representative from a small-town police department in a heavily industrialized area noted how his agency was relatively well prepared in terms of communications, hazmat, and incident command. But, in terms of PPT, he said, "We are sadly lacking." He added, "To put an officer out there with insufficient training and equipment is not right."

In sum, the low baseline of personal protection preparedness combined with high performance demands on PPT used in law enforcement creates particularly difficult hurdles standing in the way of improving the health and safety of law enforcement responders in the line of duty.

In recognizing the need to improve PPT for law enforcement, the federal government has instituted several programs, coordinated through the Office of Domestic Preparedness, to provide funding for local police departments to acquire equipment, including PPT, and to offer training. Many of the large departments with whom RAND met had taken advantage of this funding. While such programs represent a positive step toward protecting law enforcement responders, they primarily are built around responding to the threat of terrorism and are viewed as nonessential resources for large law enforcement departments with special needs. Discussion participants emphasized that widespread awareness, availability, and use of PPT still does not exist at the local police department level.

HAZARDS AND TECHNOLOGY PRIORITIES FOR LAW ENFORCEMENT RESPONDERS

Participants voiced their greatest concern about three principal threats to the safety and health of law enforcement officers: assaults, automobile accidents, and acts of terrorism. The last concern was raised by multiple emergency responder services and is addressed in Chapter Six. Other concerns that were raised include infectious diseases, nonassault injuries incurred during arrests, and exposure to chemicals involved in illegal drug manufacture. Overall, the community's perception of the risks from these threats agrees with the available surveillance data, which were presented in Chapter Two.

Protecting Against Assault

The PPT in most widespread use in law enforcement is the ballistic vest. It is designed to protect the wearer primarily from gunshot wounds, but it also provides protection from knife wounds, abrasions, and blunt-impact injuries. Two participants noted that ballistic vests helped reduce injuries from serious auto accidents, and "probably saved way more lives from blunt-force injuries from a steering wheel than they ever protected from bullets."

Nevertheless, patrol officers often do not wear vests because they find them too uncomfortable to wear over the duration of an entire shift. Problems with fit that were cited include the vests bunching up at the waist and riding up the chest and neck when seated in a car, excessive warmth, and moisture buildup underneath the vests. "It gets soaking wet all of the time," said one law enforcement representative from a warm-weather community. "Discomfort is the reason they don't wear them." Most departments that RAND visited issue vests to all responders but do not require their use for routine duties.[3]

Senior-level officers consistently mentioned the high cost of purchasing ballistic vests for their departments. A related problem is the lack of any objective means of determining when a vest needs to be replaced. Consequently, vests are usually discarded when the manufacturer's warranty expires, which is usually five years after purchase.

Research and development to improve ballistic vests, largely supported by NIJ's Office of Science and Technology, has produced tangible benefits: Several participants noted that the functionality and comfort of ballistic vests has improved significantly since they were first introduced in the mid-1970s. The thickness and weight of the vests have been reduced, but to some extent this been achieved by reducing the size of the vest, which leaves portions of the wearer's shoulders and the lower abdomen exposed. With regard to recommended technology improvements, there appears to be a strong consensus supporting development of more-comfortable vests (in terms of weight, flexibility, and breathability) with equal or improved levels of protection. Such improvements would promote greater usage of vests and allow for designs that provide greater bodily coverage.

For officers in high-risk situations, additional assault protection equipment, such as helmets, face shields, and body armor, is available. This equipment is welcome protection in predictably high-risk assignments, such as forced entry and arrest, some types of crowd control, and many SWAT missions. However, participants were in strong agreement that this type of equipment is not appropriate for officers on routine patrol, even though it is on routine patrol when the vast majority of assaults to officers occurs. A problem with this additional as-

---

[3]Ballistic vest usage was said to be in the 50–60 percent range by several participants. One explanation that was given for not making the vests mandatory is to allow families of officers who are shot and killed in the line of duty to be eligible for benefits whether or not the officer was wearing a vest. If the wearing of vests were mandatory, officers who are shot while not wearing a vest would be in violation of department policy and their families may not be eligible for benefits.

sault protection is its threatening appearance and the fact that it conveys the impression that violence is presumed to occur.

Other innovations mentioned in the discussions with law enforcement responders included "throw-on" armored jackets and overcoats that can be donned in high-risk situations (and which can also be used by fire or medical personnel), protective armor that is integrated with the uniform shirt, and armor that is worn as an outer garment with the officer's identifying information and other features sewn onto it. Participants, however, identified logistical and performance constraints that lessened the usefulness of this gear: Overcoats are not likely to be available for grabbing in a hurry; integrated-armor shirts tend to become burnished and fray too quickly, and external armor may appear threatening to the public. As one police representative put it, "As officers, you're at risk all of the time. There is no time to get a vest."

Preventing Automobile Injuries

Most police patrol work is done while officers are riding in an automobile. We often heard participants describe the patrol car as "the officer's office." Forty percent of all line-of-duty police officer deaths are motor vehicle related (see Chapter Two). About one-third of police line-of-duty fatalities and 16 percent of line-of-duty injuries result from accidents occurring while officers are in their patrol cars. Law enforcement representatives participating in the RAND discussions mentioned three major problems driving these numbers: problems with vehicle interior design; lack of protection during high-speed, rear-end collisions; and the driving behavior of patrol officers.

Vehicle interior design problems center around the location and design of a vehicle's communication and information management systems. Generally, this equipment is located to the right of the driver, and use of this equipment can cause officers to become distracted, leading to accidents. Participants also believed that this equipment increases the risk of injury in collisions because occupants can be thrown into the equipment during an accident. In addition, the size of the equipment often leaves the officer in the passenger seat with limited space, resulting in ergonomic problems. An emerging improvement, discussion participants noted, is equipment built into the vehicle itself rather than retrofitted into the passenger space. Indeed, one police department showed the RAND team a new set of patrol cars with much of the communication and information management systems integrated into the dashboard. Some participants suggested that law enforcement officers could also benefit from "heads-up" displays, similar to the displays that are used in fighter jets and some luxury automobiles, so that officers could keep their eyes on the road.

Police officers' patrol duties require them to operate and stop their vehicles in the vicinity of high-speed traffic—on city boulevards as well as on freeways. In such situations, an officer can be seriously injured or killed by a rear-end collision. One participant raised concerns about the design of police cars that contributes to fuel-tank explosions in such incidents. Options that were discussed to address the risk of rear-end collisions included strengthening the frames of police vehicles, adding active protective devices (e.g., airbags) that are specifically designed for such collisions, and improved warning lights on police vehicles. Although a few officers mentioned the concept of designing a police car from the ground up to incorporate enhanced interior and safety features (as opposed to simply modifying a civilian vehicle), none thought it was a practical alternative.

Senior police officers who discussed vehicle accidents appeared to be keenly aware that the driving behaviors of patrol officers—especially the younger members of the force—are a major cause of accidents, injuries, and deaths. Seat belt use has improved significantly in recent years, but participants reported that it is still not universal. Department rules mandating that officers pull over when using computers are not always observed. And participants said they know that officers are prone to drive at excessive speeds. Law enforcement representatives discussed various strategies that their departments have used or are considering to educate officers about the dangers of driving at excessive speeds and enforcing safer behavior behind the wheel. These measures include purchasing vehicles with lower-power engines; installing speed monitors and governors; developing protocols for high-speed pursuits, including when to desist from engaging in a chase; and disciplinary action for noncompliance with department rules. Our discussions with law enforcement representatives left us with the overall impression that many approaches are being discussed and tried in this area, but there is little knowledge about which ones are the most effective at enhancing automobile safety.

Protecting Against Pathogens

Like their colleagues in the emergency medical service, law enforcement personnel noted their concern about certain health hazards they now face in their routine duties: exposures to hepatitis, tuberculosis, and HIV. In addition to accidental exposure, many participants listed assaults, such as spitting, as potential means of exposure to pathogens. Many agencies now issue to individual officers or stock patrol cars with duffel bags or fanny packs containing disposable gloves, gowns, glasses, and masks or respirators for basic splash protection. However, one police representative from a mid-size city stated that not all personnel were trained and received refresher training in how to use the gear in the equipment bags his department issues, and he estimated that half the force did not even know what was in the kits.

Pathogen exposures through the eyes, nose, and mouth were said to account for the largest portion of workers' compensation claims in one mid-size city. At the same time, the eyes, nose, and mouth were seen as being difficult to protect in the course of routine operations because patrol officers rarely carry the necessary protective gear with them or do not have the time, or do not take the time, to return to their cars to retrieve that gear. "We're lucky to throw on latex gloves," said one officer.

> Cops use their hands for everything they do.
>
> *—Law enforcement representative*

Protection of one's hands was seen by some participants as the most difficult pathogen protection problem they face because of the need for manual dexterity in executing critical police tasks such as driving, holding a flashlight, apprehending individuals, and using weapons. The need to collect evidence also was repeatedly cited as being critical to police work, but it also makes protecting one's hands more difficult. "You have to be able to look at evidence and handle it," asserted one representative. A crime scene investigation can go on for many months, he added, pointing to the 1993 World Trade Center bombing. The anthrax attacks in 2001 also highlighted this problem: U.S. Postal Police needed to gather evidence while protecting themselves from extended exposure (Jackson et al., 2002).

Like ballistic vests, protective gloves (typically made of latex) are in widespread but not universal use in law enforcement. Most law enforcement agencies reported having gloves stocked somewhere in their patrol cars. However, the protective capacity of latex gloves is limited. Police officers worry about needle sticks and other hand injuries when searching pockets or cars for evidence. Leather gloves (sometimes lined or coated to prevent needle-sticks) were seen as adding a level of protection, and they appeal to many officers (partly because of the "macho factor," said one participant). However, it was noted that an approaching police officer wearing leather or even latex gloves can be perceived by the public as intimidating and may communicate undesirable and unintended messages.

Chapter Six

PROTECTING HAZMAT AND ANTI-TERRORISM RESPONDERS

In the wake of the September 11, 2001, attacks, protection from hazards associated with terrorism response has become a high priority for the entire emergency responder community. For many of the discussion participants, September 11 provided a graphic example of the wide range of hazards and protection needs associated with terrorism response.

Most large fire departments that participated in this study expressed confidence in their ability to respond to typical hazardous materials (hazmat) events, such as hazardous cargo spills from trucks involved in highway accidents. In contrast, these same departments were greatly concerned about their ability to respond to large or multiple acts of terrorism. For instance, terrorism response could require participation of rank-and-file first responders who are typically underequipped and undertrained for such activities. In addition, response to an act of terrorism is likely to unfold more quickly, in more unexpected locations, and with much more uncertainty than response to an industrial hazmat incident. Concern about protection needs for more conventional[1] hazardous materials response was also raised in the discussions, although it did not emerge as an area of major concern for most of the emergency responder departments we visited.[2]

This chapter addresses a limited set of shortcomings associated with protective gear for conventional hazmat response and major concerns over protection when dealing with a terrorist attack that may employ chemical, biological, radiological, or other hazardous materials. Identifying and monitoring hazardous materials were also mentioned as being important components to protecting

---

[1]In this report, conventional hazmat response refers to situations in which specially trained and equipped hazmat technicians are deployed in response to an initial assessment made by personnel at an incident scene. It is not normally part of the first response.

[2]As discussed in Chapter One, the sample of departments we used is biased toward larger departments relative to the national average.

emergency workers who are responding to a hazmat or terrorist event. This topic is further addressed in Chapter Seven.

CONVENTIONAL HAZARDOUS MATERIALS RESPONSE PROTECTION

Twenty-five of the 33 fire departments included in this study provide hazmat response capability. This finding follows an observation that arose in our discussions with responders—hazmat demands are increasing. As Figure 2.3 illustrated, there was an 86 percent increase in the number of fire department hazmat responses between 1986 and 2000. Participants also noted that hazmat response is becoming more diverse in the types of materials and situations that are involved, and it is no longer solely the domain of large fire departments.

As economic development and the presence of hazardous materials become more widespread in the United States (e.g., many rural communities now have industrial facilities or have a highway, rail line, or pipeline passing through their jurisdictions), more medium- and small-size departments are instituting hazmat response capabilities in response to emerging local needs. One participant drew an analogy to emergency medical service response: Just as the increase in medical calls has led many departments to increase the number of emergency medical service responders and provide EMS response capability from multiple stations, the emergency responder community may witness an increase in the number of hazmat responders and an expansion of their emergency response role.[3]

> Our mission has gone from "Put the wet stuff on the red stuff" to making our communities a safe place. We are doing many more things, and it is becoming more technologically demanding.
>
> —*Fire service leader*

Despite the trend toward increased hazmat calls, many fire departments do not have a hazmat response capability. Instead, they rely on a neighboring locality's capabilities. Community views on the merits of this approach varied. Some departments were satisfied with such arrangements, while others felt that more local capabilities were necessary. A participant from one such department expressed grave concerns regarding the low level of hazmat training for the firefighters in his department and the insufficient number of trained personnel regionally available to respond to a large industrial or transportation-related hazmat event.

---

[3]In several fire departments that participated in the study, all firefighters were also certified EMTs. In contrast, in only one department was every firefighter also a hazmat technician.

In assessing hazmat protection, a common theme in the discussions was that protection for conventional hazmat response was generally very good. Many participants attributed the quality of protection to the fact that, compared with firefighting, most hazmat response employs a less-urgent and more-methodical protocol-driven approach. This difference between firefighting and hazmat response may be because rescue is less often a component of hazmat response, and because hazmat response is newer and less tradition-bound than firefighting. As a result, personal protection technologies are designed with high levels of protection and training in mind. Ancillary gear requirements, such as pockets, the ability to be donned rapidly, and allowing for a high degree of agility are less of a priority. Nonetheless, some shortcomings with protection for conventional hazmat response were identified.

One concern that participants highlighted was the need for protection against substances or combinations of substances that present multiple hazards (i.e., "multicharacteristic" protection). Hazmat personal protection is designed primarily for chemical protection, but in many instances, the wearer is exposed to other hazards as well, the most common one being fire. Hazmat spills often occur as the result of vehicle collisions or structural collapse, in which the potential for fire may be high, and many hazardous chemicals are highly flammable. Referring to its poor flame resistance, one participant called a Level-A suit[4] a "body-bag with a window." While current standards do require hazmat suits to have a certain level of flash protection, manufacturers typically meet these standards by providing a flame-resistant overcover, a solution that was generally viewed as being suboptimal by the hazmat responders with whom we met. Other combinations of hazards related to terrorism for which adequate protection is a concern are discussed in the next section.

Secondary concerns among some departments were the difficulty in reusing hazmat suits and the uncertainty surrounding that reuse. The difficulty arises from the need to decontaminate, pressure test, and visually inspect suits after each use, and the uncertainty arises from lingering doubts about whether the suits are really clean. These concerns, along with the high cost of reusable hazmat suits, have resulted in a nearly wholesale shift in the emergency response community to inexpensive, "limited use" (typically interpreted to mean "disposable") hazmat suits.

---

[4]The Environmental Protection Agency classifies four levels of protective clothing ensembles to be worn when dealing with hazardous materials. A *Level-A suit* fully encapsulates the body so that no vapor penetrates the suit; respiratory protection is provided through supplied air (such as an SCBA). A *Level-B suit* is a full-body chemical-resistant suit that may introduce vapors; respiratory protection and other protection features are normally the same as with a Level-A suit. A *Level-C suit* is a full-body chemical suit with the same properties as a Level-B suit, except that an air-purifying respirator is used instead of supplied-air respiratory protection. A *Level-D* suit protects against contact exposure only, and no respiratory protection is required.

TERRORISM PROTECTION

A recurrent major concern among emergency responders was protection from chemical, biological, radiological, and conventional explosive terrorism. The threat of terrorism directed toward U.S. citizens has heightened emergency responders' awareness of the risks and hazards involved in responding to such incidents. This concern transcends service divisions and represents one of the largest gaps between perceived hazards and available protection we discovered during our discussions with emergency responders. The majority of responders feel vastly underprotected against the consequences of chemical, biological, or radiological terrorist attacks.

Shortcomings with Conventional PPT

> WMD has dramatically changed how bomb technicians work.
>
> —*Bomb squad representative*

While many municipalities have separate hazmat, bomb, and SWAT response capabilities, these resources could be quickly overwhelmed after a terrorist attack. In the past, a terrorist threat was typically conceived as a static event—for example, a package containing an explosive device delivered to a target location—for which bomb technicians have well-developed safety protocols and equipment that emphasize defensive tactics. In such cases, remote-controlled robots and disrupters (bomb deactivation devices) have protected responders by enabling them to avoid having to handle dangerous materials.

Participants also expressed concern about having adequate protection against new combinations of hazards and about responder roles that are emerging as a result of the terrorist threat. For example, there is rising concern about chemical or "dirty" bombs, in which an explosive is used to disperse a chemical or radiological agent. To prevent the release of WMD agents while dealing with such devices requires a return to the "live entry" and manual disarming practices of the past, which then requires a combination of chemical and explosive protection. Similarly, the increasing role of the "human element" in crime and terrorism, such as heavily armed assailants or terrorists seeking to create maximum impact, may require responders to use aggressive intervention tactics. These more-aggressive tactics may require ensembles with chemical, assault, and ballistic protection that still allow for vigorous physical activity. Protecting responders in these cases requires marrying hazmat and SWAT or bomb squad protective equipment. For example, conventional bomb suits have been modified to be worn over a Level-B hazmat suit in conjunction with an SCBA. Several of the bomb technicians we interviewed expressed the concern that this solution compromises the suit's protection against explosives and limits the wearer's mobility when it might be needed most.

Many participants also noted that hazmat gear is not designed for extended or repeated use, which would likely be the case with a WMD event. Chemical protective suits tear easily, and protective equipment degrades with repeated decontamination.

Finally, hazmat gear traditionally has been rarely used, and, thus, is not stockpiled in significant quantities. "Most departments are not equipped to deal with several hundred contaminated victims," said one fire service official.

Chemical Protection Needs of Front-Line Responders

Participants from fire, police, and emergency medical services anticipated that they would play important roles in a terrorist-type event. Police noted that they would likely be the first on the scene, would execute enforcement activities if terrorists are present, would preserve the essential characteristics of the crime scene, and would manage the perimeter of the incident site, including facilitating the removal of injured or contaminated persons. Firefighters would have primary responsibility for containing hazardous substances, suppressing fires, and conducting search and rescue for victims. Emergency medical technicians and paramedics would likely assist with rescue operations and render medical care to injured individuals.

Because of the potential for rapid onset and large scale with such events, those activities may have to be carried out by front-line responders rather than specialized units. Consequently, many of the emergency response departments expressed a strong desire for chemical protection for the "regular" nonspecialist firefighter, patrol officer, paramedic, or EMT. Prompted by the increased threat of terrorism, available federal grant money, a strong sense of urgency to be fully prepared for response to terrorist attacks, and, in some cases, experience in responding to terrorism, several fire and police departments were considering or, in several cases, had acquired chemical and respiratory protection for all of their emergency responders.

> We are working toward respiratory protection for every police officer on the street.
>
> —*Law enforcement representative*

After the problems with inadequate personal protection that were experienced in the aftermath of the 1995 attack on the Murrah Federal Building in Oklahoma City and the September 11, 2001, attacks, adequate chemical and respiratory protection for law enforcement is now seen as a critical component of proper scene control (Jackson et al., 2002). Accordingly, several police departments indicated that they are equipping squad cars with chemical-protective gloves, suits, escape hoods, and respirators. Similarly, several fire departments currently stock some sort of chemical-protective suit on all apparatuses. A manufacturer noted that sales of chemical-protective suits to the (nonhazmat) fire service were on the rise. A few fire departments reported that they have also acquired emergency escape hoods. For every department that had acquired such protection, another was actively pursuing the same protection.

This trend toward increased chemical and respiratory protection reflects an important addition to the continually expanding role of the local emergency responder. This additional role has fundamental implications for individual responders in terms of how they view their responsibilities and for departments in terms of equipment and training needs, operational procedures, and regional mutual-aid agreements. This added role also has important implications for the federal government and other agencies that play a part in researching, guiding, and overseeing the development and implementation of personal protection for emergency responders. These implications include the need to reexamine the protective requirements and users' operational needs in the design of protective equipment, operational protocols, training programs, and interagency coordination for responding to terrorist attacks.

Uncertainties Surrounding Chemical Protection

> Departments don't understand what [chemical protective clothing] is good for. They have a false sense of security.
>
> —PPT supplier

One of the most significant findings from our discussions with the responder community on the topic of hazmat and terrorism protection is that departments are proceeding down the path of acquiring chemical and respiratory protection without having a clear understanding of what exactly they are preparing for and how to prepare for it. In some cases, participants admitted to these uncertainties but nonetheless felt that they could not wait for them to be sorted out before acting. Our discussions revealed that the issue of providing protection for chemical, biological, or radiological (CBR) terrorism is fraught with several uncertainties, including the following:

- The nature of the threats and hazards that emergency responders will face

- The types of protection that are appropriate and how to obtain them

- Whether the protective technologies will work

- How these technologies will be integrated into operational procedures.

The range of potential threats and event scenarios is vast. While departments in some regions of the country have conducted generalized threat assessments, few of the departments have assessed in much detail the anticipated hazards to which emergency responders would be exposed. As a result, there is considerable uncertainty about the hazards that emergency responders should be prepared to face after a terrorist attack. Such uncertainty frustrates efforts to design a protection program and acquire the necessary technology to support that program. Law enforcement departments, in particular, said they did not know what they should be protecting against, what level of protection was appropriate, or where to look for that appropriate protection.

Another fundamental uncertainty revolves around how well the available protective technologies will work for the anticipated situations. While hazmat protection is subject to rigorous standards and certification procedures, those requirements are designed primarily around the conventional hazmat response model.[5] As a result, participants noted, much of the available hazmat protection is neither designed nor certified for this new role of CBR terrorism response. Thus, departments are looking outside the traditional supply channels to locate the appropriate equipment. Several departments felt compelled to forge ahead alone and use whatever guidance they could get to make purchasing decisions. For example, some departments were acquiring equipment based on standards from the International Standards Organization (ISO) and North Atlantic Treaty Organization (NATO) and other military standards, even though those standards do not apply to municipal emergency response departments in the United States. In other cases, departments were using personal connections to obtain unofficial equipment-performance evaluations that manufacturers have conducted but are unable or unwilling to publish. Several participants expressed frustration over the lack of appropriate standards and guidance in the area of CBR terrorism protection.

In response to some of these concerns, the NFPA has recently developed a protective clothing standard for chemical and biological warfare agents. This standard specifies three levels of protection based upon the ability of a protective garment to resist various substances and the durability of materials used in the garment. The goal of this standard is to establish personal protection requirements for ensembles that would (1) be available in quantity; (2) be in pristine condition; (3) be designed for single-exposure use; and (4) be easily donned and used by fire and emergency services personnel to reduce the

---

[5]The Environmental Protection Agency hazardous materials protective clothing classifications (Levels A-D) were defined primarily for workers at hazardous waste sites, where emergency conditions typically do not exist.

safety and health risks to those personnel during assessment, extrication, rescue, triage, and treatment operations at or involving chemical or biological terrorism incidents (National Fire Protection Association, 2001b). Because this standard was recently introduced, the emergency responder community may not have had a chance to fully evaluate it in the field.

Similarly, NIOSH has begun issuing respiratory protection standards for chemical, biological, radiological, and nuclear warfare agents. The SCBA standard was finalized in December 2001, and the first compliant SCBA was certified in June 2002. As with the NFPA chemical protection clothing standard, the influence of this standard on the emergency responder community may not be felt for some time. Standards for APRs and emergency escape hoods are currently under development.

Also unclear is how this protective technology is expected to be used. Speaking of their department's response to anthrax calls in the autumn of 2001, one fire service leader commented, "We had 100 different approaches to these incidents." A number of questions regarding this issue were raised in RAND's discussions with participants:

- When is personal protective equipment to be used?

- Should personal protective equipment be used for operational purposes or for escape only?

- What activities do emergency responders envision conducting in a CBR event?

- Who will make these operational decisions?

For example, many emergency responder departments are equipping their vehicles with escape hoods that enable a responder to exit dangerous environments. However, there is currently much confusion over how a responder will know when to don the hood. In addition, if the escape hood is stored in a vehicle and the responder is away from the vehicle when a dangerous situation becomes apparent, the responder cannot access the hood. Law enforcement representatives also wondered if escape hoods would be used by police officers for operational purposes (e.g., rescues) rather than solely for exiting a hazard zone. In another example, an EMS representative noted that his service had equipped its trucks with chemical-weapon antidote auto-injector kits at great cost. However, he also raised concerns about their use and potential misuse: What if, he proposed, "You go to an incident and you think a nerve agent [is present]?" Responders will be tempted to use the kits at times when it isn't necessary, he surmised. "It's a problem."

Technically, Level-A protection is required for an unknown hazmat environment. But we still need to do our job, and there's no way that we can do that in a Level-A suit.

—Law enforcement representative

When considering the sort of activities emergency responders might conduct after a terrorist attack involving weapons of mass destruction, limitations arise from existing hazmat protection having been designed around the conventional hazmat response model. For example, firefighters cannot fight fires using current methods while wearing chemical-protective suits. Law enforcement officers, in particular, need to be able to run, access their weapons and other tools quickly, engage in covert operations, and make arrests, none of which is feasible while wearing the currently available hazmat protection equipment. Speaking of his agency's efforts to acquire Level-C protective equipment for its officers, one police department leader stated, "Vendors have not come up with anything that is appropriate for law enforcement."

An additional complication with existing and emerging chemical protection is the difficulty with logistics for specialized personal protective equipment. Many protective equipment components are assigned to vehicles rather than issued to individuals. The main reason is cost.[6] For example, a firefighting apparatus may carry four chemical protective suits that are available to whomever happens to be riding on the apparatus. As a result, there is no way to ensure that the correct size suit is available for each person. Some departments address this problem simply by stocking nothing but extra-large sizes, while others stock a range of sizes. In any event, improper fit can affect both the effectiveness of the chemical protection as well as the wearer's ability to fully function.

Most departments have not yet had an occasion to use their chemical-protective gear in an actual emergency, so the extent of the problems outlined above remain largely unknown. However, it is clear that in many cases there is a substantial gap between the perceived hazard and a clear understanding of the appropriate personal protective equipment and practices for dealing with that hazard.

---

[6]Equipment assigned to vehicles can be shared by all shifts, requiring less inventory. Also, respiratory protection assigned to vehicles for "emergency only" is often exempt from the costs of complying with a regulatory requirement dictating that equipment assigned to individuals can be used only after that individual satisfies a pulmonary fitness requirement and performs a fit test.

Chemical Protection Challenges and Alternatives

While several of the larger departments are acquiring chemical and respiratory protection, many departments reported that they were struggling to provide any such protection. Impediments to providing the protection include the additional equipment costs and the difficulty in dedicating the staff and time to conducting pulmonary physicals, fit testing, and training, especially in realistic scenarios. In some law enforcement departments, the only forms of respiratory protection they had available for their personnel were Vietnam War–era surplus military gas masks, which, according to one participant, were "usable, but for [protecting against] tear gas only, and some don't work." In other departments, the respiratory protection that was made available to police officers consisted of hand-me-downs from the local fire department.

Given the lack of respiratory protection, many participants mentioned the need to place greater emphasis on precautionary measures before entering the scene of an incident, such as identifying hazardous agents, determining wind direction and plume behavior, and assessing risks. Many participants also expressed the opinion that setting up incident staging areas in safe locations was imperative, and that first responders in particular were getting better at doing this. However, without specialized training, there is a limit to the ability of nonspecialist responders to take appropriate actions, even for basic functions such as hazard awareness. This is especially true in law enforcement, which one representative of big-city agency described as having no hazard awareness and no hazardous material operational training.

Chapter Seven

SYSTEMS-LEVEL PROTECTION ISSUES

The previous four chapters focused principally on individual-level protection for personnel in different emergency response organizations who face a range of diverse hazards. In this chapter, we broaden our scope and examine protection issues at the systems level. *Systems-level protection* refers to protective technologies that operate at the command or unit level and include communications, hazard monitoring and assessment, personnel management, and various "human factors." The difference between the two levels of analysis may be difficult to define precisely, but the distinction carries some important conceptual implications. In particular, addressing systems-level issues is likely to be more complex and involve more stakeholders than addressing individual-level issues, but systems-level technologies also have the potential to have a greater effect in terms of meeting protection needs.

COMMUNICATIONS

The need for better communications was a universal theme heard in the RAND discussions. This need is driven by the desire to improve the management and safety of personnel as emergency response incidents become more complex. Moreover, information and knowledge—gathered and shared via communications networks—are becoming more critical to decisionmaking and safety.

Tactical Communications

Firefighter representatives were particularly concerned about shortcomings in existing tactical communications technologies given the conditions that exist in the environments in which they work, such as high ambient noise levels and intense heat, and the difficulties in communicating through a respirator. (These issues are discussed in Chapter Three.) However, these concerns are increasingly salient to emergency medical service personnel, law enforcement personnel, and other responders because, as we have seen, many departments are acquiring SCBAs and other technologies to protect their personnel against terrorist and other threats. Readers interested in personal communications are therefore encouraged to refer to the discussion in Chapter Three.

Strategic Communications

Both police and fire departments emphasized strongly that there are fundamental problems with radio communication systems that extend beyond the tactical problems just mentioned. These problems have to do with networks and protocols governing communication among individual responders, departments, and services. As such, they transcend the boundaries between these organizational elements and therefore are strategic concerns for the entire emergency responder community.

One problem with radio communications systems that was cited in the RAND discussions is that police, fire, and emergency medical services in many jurisdictions use incompatible radio systems and therefore cannot communicate easily with each other at incidents. A lack of interagency communications has been cited as contributing to the lack of coordination between the New York City police and fire departments at the time of the imminent collapse of the World Trade Center towers on September 11, 2001 (McKinsey & Company, 2002). Interagency communications are a critical enabler of the Incident Command System (ICS)[1]: "We collocate on a regular basis, but we don't have the ability to communicate," said one big-city law enforcement representative.

A related concern is the problem with interjurisdictional communications: Departments in the same service from different jurisdictions often are unable to communicate, leading to coordination problems in mutual-aid situations. One participant related that after a massive tornado struck his city in the late 1990s, the incompatibility of communications systems among the region's jurisdictions prevented authorities from coordinating their activities. The result was a "nightmare": Too many units arrived at the scene, many of which were self-dispatched, hampering response efforts.

Similarly, participants spoke of incompatible communications systems among local, state, and federal agencies, a growing concern in a period of heightened awareness about terrorism threats. In one example, a representative from a fire department whose jurisdiction neighbors state wildlands and hence routinely engages in joint operations with state forest firefighters, noted that the two

---

[1] The *Incident Command System* is a standardized approach for organizing and managing emergency responses at incident scenes. The ICS management structure consists of five major components: the incident command (including a command staff), operations, planning, logistics, and finance/administration. The ICS includes a common terminology to allow interagency communication, standardized organizational processes, and a scalable incident management structure.

groups could not communicate with each other because they used separate radio frequencies.

In response to these problems, there has been a strong push in recent years in many communities to modernize communications systems. In an effort to ensure reliable interservice and interjurisdiction communications, communities are transitioning from conventional analog radio-to-radio technologies to higher-frequency (800 megahertz [MHz]) "backbone" or "trunked" networks that rely on a system of base stations and repeaters permanently installed in a service area that manage and relay radio signals. These efforts have been supported by grants from the federal government and state and regional emergency management authorities. The primary drivers of implementing trunked systems are that the systems allow all users to intercommunicate and they can be scaled up to accommodate additional services and users over time. The systems also provide enhanced transmission clarity in most environments, employ automated frequency control to help manage radio traffic, and monitor radio users' identifications to facilitate personnel accountability.

We found in our discussions, however, that many departments that had acquired these systems were not fully satisfied with their performance. Participants cited both technological and organizational problems. One complaint about the trunked systems involved constraints on communication behaviors. Some participants disliked the 1.5-second pause the repeater generates when a user activates the transmission switch, which often clips off messages. Others complained that the system does not allow users to talk over one another. If a responder overmodulates or forgets to release the transmission switch in a stressful or panic situation, others cannot interrupt or speak over that responder. Some participants noted that working with these limitations could be addressed through training and experience with the systems, and that problems would diminish as users became more familiar with the technology.

> This [800 MHz trunked] system doesn't work. It needs too many repeaters, which costs too much money. The idea is good, but it's too expensive to implement.
>
> —*Fire service representative*

One of the main technological problems cited in the discussions was unreliable signal transmission. Signal loss and resulting "dead spots" were said to be most problematic in areas with tall buildings or hills, and particularly in areas below grade, such as basements and parking garages. The limited signal penetration into and within high-rise buildings and other difficult environments was seen as being comparable to, if not worse than, the signal penetration with analog systems.[2] Non-line-of-sight and intrabuilding transmission problems can be improved by increasing the number of repeaters supporting the system. Buildings can also be outfitted with "leaky feeder" systems: cables routed throughout a structure that act as an antenna. However, participants stressed that these solutions added significantly to system cost and, as one fire service representative noted, "most municipalities can't afford this." The situation is complicated by questions about who is responsible for installing and maintaining such networks: Are these tasks the responsibility of emergency response agencies, the municipality, or building operators?

Beyond technical considerations, cost and coordination issues are also serious impediments to widespread implementation. Even with financial support from the federal government, the need to coordinate policy and acquisitions across many agencies slows implementation. After seven years of discussions, 14 agencies across one mid-size metropolitan region were still two years away from system implementation. "There does not seem to be a lot of enthusiasm to participate," said one representative. Other participants pointed out that even when a system is adopted, the system is underfunded, or not all agencies opt to participate, thereby reducing potential system effectiveness. "They bought a $10 million system and put $2 million into it," said one representative from a mid-size city. Of particular concern was a lack of participation by state and federal agencies, such as the National Guard, Coast Guard, Federal Bureau of Investigation, and U.S. Drug Enforcement Agency.

Yet another challenge with high-frequency trunked radio systems that participants noted is the unresolved problems with frequency allocation and resulting interference among public safety, private wireless services, and commercial, industrial, transportation, and specialized mobile radio users. These problems must be sorted out, participants claimed, before communitywide investment in and implementation and acceptance of a single interoperable communication system are possible.

---

[2]With analog radio-to-radio systems, signal loss can be overcome in some situations by "talking around" the problem, i.e., having personnel nearby relay messages along a chain of individuals. Because transmissions in trunked systems are relayed by repeaters, a user out of range of a repeater has no capability of contact with others. Such concerns led the New York City Fire Department to abandon use of a trunked system after testing it on a pilot basis (Dwyer, Flynn, and Fessenden, 2002).

A unit commander will sometimes need to be operating two radios: An analog system to communicate with his unit inside the building and a trunked system to communicate on the strategic channel.

—Fire service representative

Given the shortcomings of existing communication systems and the slow pace at which trunked systems are being implemented, few options for interoperable communications exist. Many departments resort to using several different systems to handle all of their communications needs. Unit commanders often use conventional analog radio-to-radio systems to reach responders inside buildings and digital trunked systems to communicate with other services, jurisdictions, and agencies. One agency in a small county reported that their emergency operations truck incorporated seven distinct communications systems: two analog radios, an 800-MHz radio, a cell phone, a pager, a marine band radio, and a satellite phone. Another alternative mentioned in the discussions is that responders sometimes resort to commercial mobile phones and pagers. Again, responders noted shortcomings with this backup option: In large-scale incidents such as natural disasters, industrial accidents, or terrorist attacks, mobile networks can be overwhelmed by heavy civilian use. To solve this problem, several participants called for requiring commercial mobile service providers to give precedence to designated agencies for airtime in certain emergency situations.

HAZARD ASSESSMENT

An important aspect of protecting the health and safety of emergency responders is the ability to detect, monitor, and assess an environment for thermal, chemical, structural, explosive, and other hazards. These tasks help responders decide how to approach a situation and what types of personal protection they should use. Hazard detection, monitoring, and assessment take place in some form at every incident. The tool most commonly used by emergency responders is personal experience. "We go on our knowledge," said one firefighter. This knowledge may include a firefighter's experience with how a fire progresses or a police officer's experience with a particular individual or location. Emergency responders also frequently resort to simple indicators and rules of thumb, such as those concerning the characteristics of a smoke plume or human behavior.

While such approaches are often indispensable, participants in the RAND discussions noted that these approaches also are often inaccurate and insufficient given the increasingly complex and uncertain environments in which emergency responders must operate. For example, building construction and building materials are evolving rapidly, and firefighters' experience with actual fire situations is decreasing as a result of the long-term decline in the number of structure fires.

To improve the accuracy and usefulness of hazard assessment approaches, RAND participants suggested improvements in information availability, monitoring technologies, and assessment tools that would help them better understand the hazards they face in the line of duty. Dealing with emerging hazards associated with terrorist events, highlighted by the anthrax attacks of 2001, were cited as a particularly important concern.

Hazard Information

Information available on site is increasingly being used to assess hazards. Many firefighters commented on the value of placards on buildings and vehicles indicating the presence of flammable, reactive, toxic, caustic, explosive, or otherwise hazardous materials. While useful when it is available, participants noted that such information often is not posted or regularly updated, even when required by code. Given the proliferation of building materials and construction types, several firefighters expressed a desire for a similar type of placard system to identify building design. Fire service representatives repeatedly mentioned the increased building collapse hazard in buildings with lightweight truss roof and floor construction, which is now used widely in commercial structures but is often not readily apparent to responders.

Another informational tool used in emergency response is a "pre-plan," which includes information compiled in advance on, for example, hydrant and standpipe locations, utilities, building design and layout, hazardous material inventories, and service histories from previous calls. Pre-plans may be developed by municipal services or by industries to guide their emergency response personnel. "It will change the way you will attack the fire," said one participant. Industry representatives expressed satisfaction with their plans. The usefulness of municipal pre-plans, however, was questioned. Some municipal representatives noted that pre-plan information often is stored in a format and location that are difficult to access (e.g., paper copies stored in the fire chief's vehicle). As one firefighter noted, even when pre-plans exist, "In reality we don't usually have that information [on hand]."

In addition to on-site information and pre-plans, a third type of hazard information is provided by dispatchers, who gather information from callers or other personnel on the scene. Pre-plan and dispatch information could be made much more usable, participants claimed, by exploiting information technology. Examples of emerging capabilities that are beginning to improve information utility include computer-aided dispatch involving the transmission of dispatch information to mobile data terminals in emergency response vehicles and integrated Geographical Information System and Global Positioning System (GPS) technologies that generate maps and floor plans, guide vehicles, and locate critical items such as hydrants, stand pipes, and hazardous materials.

Environmental Monitoring Equipment

> We all need better detection equipment so we know what we are dealing with.
>
> *—Fire service representative*

Portable environmental sensors and analytic devices are coming into more widespread use in the emergency response community to assist the community in its approach strategies and PPT decisionmaking. While most of these devices are used primarily by specialized hazmat teams, participants noted that as technologies improve and become easier to use and prices drop, detectors are increasingly making their way into initial response efforts.

> Representatives from one fire department ticked off their priorities for improving infrared imagers. Based on community input, these requests can be extended to many other sophisticated personal protective technologies.
>
> —For it to be acceptable, it has to be lightweight. Affordability is second.
>
> —It's got to be fast [to set up].
>
> —It has to be easily maintainable.

One tool becoming increasingly commonplace in the fire service is the infrared thermal-imaging camera, with one-fourth of fire departments currently using this technology (U.S. Fire Administration and National Fire Protection Association, 2002). According to study participants, thermal imagery is used mostly for identifying hot spots and determining building integrity during over-haul. It is also used in wildlands fires to identify hot spots in vegetation and root systems. "We are making a lot of decisions based on thermal images," said a fire department representative from a mid-size city. In principal, thermal imagery can also be used to locate fallen personnel, and one representative relayed a case of using thermal imaging at the scene of an automobile accident to locate and recover a severed limb, which was later successfully reattached.

Infrared imaging technology has improved significantly since it was first introduced in the 1980s. Reductions in equipment size and price have occurred, but the lightest palm-size versions were still seen as an expensive option by many departments. One department in a mid-size city had acquired eight older-model imagers and was seeking more to outfit every apparatus in the service.

The State of California is planning to provide all services in the state with thermal imagers. Emerging video uplink capabilities that can transmit thermal images to commanders were seen as a promising tool for monitoring personnel and operations. Integrated "heads-up" thermal-image displays built into respirator face masks, similar to those developed for military applications, were also seen as a promising development.

First-responding firefighters typically are equipped with a basic four-gas[3] monitor, which is used primarily to determine carbon monoxide levels during the overhaul phase (to ascertain when personnel need to wear SCBAs) and to investigate "strange odor" calls, such as those for natural gas leaks. One big-city fire department reported having placed four-gas monitors on all of its ladder trucks by the mid-1990s. A small-city department reported that all three of its fire squadrons as well as its hazmat team had them.

> Right now a [chemical] hazard assessment is not done at "regular" fires. It should be.
>
> —*Fire service representative*

Many agencies participating in the RAND study reported that they have been acquiring an array of portable devices designed to sample and analyze a variety of gases, liquids, and solids. Many of the firefighters felt that chemical hazard monitoring should be routine at all fires, given the increasingly exotic and often unknown materials encountered in industrial operations and in building construction and interiors. Police and fire departments alike expressed a strong desire for improved sensing capabilities for first responders, in particular to warn responders of an unrecognized terrorist or other hazmat exposure risk. For example, one department reported that it is now seeking to place radiation dosimeters on every fire apparatus. "It's a key item for use, absolutely," said a department representative.

More advanced devices, such as portable gas chromatographs or infrared spectrometers, however, are confined to the specialized hazmat teams in most communities and are used only at incidents in which hazardous materials are known or suspected to be present. This restricted use is driven by the fact that the gear is bulky, takes time to set up, and requires special training to use. Equipment cost also was cited as a significant constraint on its wider distribution and use. As an EMS representative noted, "I can't buy enough. I can't

---

[3]Such devices typically measure the amount of oxygen, carbon monoxide, hydrogen sulfide, and combustible gases.

afford it." First-responding police officers typically have no sensors at all at their disposal.

To help prevent first responders from walking into a hazardous zone unknowingly, several participants recommended the development and diffusion of passive "badge-type" chemical and biological detectors worn on garments, similar to radiation badges. Many also wanted long-range, high-sensitivity "prior to lethal" detectors that could provide information about environments that are dangerous to life and health before responders enter those environments. The large number of anthrax scares in autumn 2001 has given rise to the desire for quick, easy-to-use test kits and monitoring devices to detect and identify chemical and biological substances, similar to those used for identifying illegal drugs. Not only can such technologies guide PPT use, but better on-site information can also reduce unnecessary equipment use and decontamination needs. A clear understanding of the risks at hand also can "mitigate the fear factor" among responders and the public, commented one participant. Because of the increasing diversity of potential hazards, stated many respondents, ideal monitoring and identification technologies must provide comprehensive solutions—i.e., those technologies must have broad-spectrum detection capabilities so that multiple instruments are not required.

In the context of large-scale structural fires as well as emerging threats of terrorism, several departments suggested that chemical sensors would be particularly valuable when installed as permanent fixtures in buildings. Broadly analogous to the benefits of smoke-detection and fire-sprinkler systems, participants envisioned chemical sensors that could trigger warning alarms and automatically implement mitigative actions. For instance, air-handling systems could be manipulated to move smoke or chemical hazards away from building occupants and responders. Chemical and radiation detection systems are in widespread use today in industrial facilities; therefore, it was argued, they could easily be developed for public buildings. Such sensor information would be yet more valuable, according to many participants, if it were available to emergency responders at the station or en route to the scene. Commercial systems that can transmit fire panel information, such as temperature sensor readings and the location of activated alarms and sprinklers, directly to responder vehicles are becoming available. This existing technology could be readily adapted, participants suggested, to also convey additional chemical, biological, and radiological sensor data.

> Chemical sensors are not that useful, especially in the initial response, because you need to be in the hazard to make the measurement, so you already need the protection.
>
> —*Hazardous materials specialist*

Some participants questioned the merit of substantial investments in hazard indicators and monitoring. One department argued that much of the discussion about improving responder protection through better environmental hazard monitoring may be misguided because an emergency responder in principle should be wearing adequate protection when initially sampling an environment with suspected hazards. Similarly, participants argued that basing protection decisions on a reading for a restricted range of potential hazards may be dangerous. For example, it was argued that carbon monoxide monitoring alone during overhaul was insufficient: SCBAs should always be worn during overhaul because carcinogens released from building materials and building contents may be present but not readily measurable.

Some fire service representatives also noted that the incremental value of improved environmental hazard monitoring was limited given existing policies in the fire service that allow for only a single PPT option (full bunker gear and SCBA). As a result, most fire service representatives stated, hazard information is more often used to guide operational decisions than to influence personal protection selection because responders must default to maximum protection regardless of the level of environmental exposure. In the case of law enforcement personnel, they typically have so little in the way of personal protection equipment and training that the issue of complex monitoring is largely moot.

An additional issue surrounding the use of environmental monitoring technologies is the level of confidence in their reliability. Several fire service representatives commented on the unreliability of warning indicators, such as personal-alert safety system alarms that signal when a firefighter may have stopped moving. Frequent false alarms, it was argued, motivated many firefighters to not activate those devices. "Smart-ticket" and other technologies used to detect and identify agents such as anthrax and Ricin must also be more accurate than they are now. A false positive, it was noted by two participants, can lead to unnecessary panic among the public as well as among emergency responders, and "is doing the terrorist's job." False negatives are even more deleterious because they may motivate responders to operate in unsafe environments. Such concerns typify the challenges involved in introducing new technologies in general, and are amplified in this arena given that the safety and health of emergency responders is at stake.

LOCATION TRACKING

Police and fire department representatives also expressed a strong desire to acquire technologies to monitor the location of individual responders. Participants stated that the primary application of such a capability would be to quickly locate a trapped or injured responder. Additional applications include managing operations at large incidents, such as natural disasters, and guiding personnel through buildings for the purposes of escape, pursuit, or locating spots of concern. Eventually, participants envisioned, simple robots equipped with location monitors and cameras could be dispatched to generate maps of incident scenes that could be used to guide responders. Location tracking of vehicles can facilitate more efficient dispatching and help to manage or investigate driving behavior.

> Location monitoring would be great for rescue
>
> —Fire service representative

Participants noted that personal location technologies based on the Global Positioning System are becoming available. In one major city, all fire apparatuses are outfitted with GPS that can be monitored by dispatch. "Now we want to take this to the people [level]," said one participant. "I am interested in a GPS locator so I can tell where the [fallen] guy is," stated one leader of a big-city fire department. Participants frequently cited the use of personal location technologies in the military, but they are seen as being prohibitively expensive for municipal use. GPS also suffers from poor vertical resolution and signal penetration problems in large or underground structures, limiting its applicability in multistory buildings.

Other location technologies are in discussion and development, it was noted, such as technologies that employ radio triangulation (exploiting differences in travel times of radio signals between a source and multiple receivers), radar (exploiting the travel time of reflected radio signals), and inertial tracking systems (using accelerometers to compute cumulative movement, also known as "dead-reckoning" systems). Radio-based systems may require either fixed antennas to be installed in buildings or temporary beacons to be placed around the site—raising cost issues not unlike those with trunked radio systems. Inertial systems suffer from cumulative drift errors that limit their usefulness. Hybrid systems are also being researched: One novel concept employs GPS-equipped vehicles parked around an incident scene to act as position beacons for radio location systems.

Should the cost and technology hurdles be overcome, two participants envisioned that individual-level location tracking technologies could be linked with communications systems and monitoring technologies to transmit emergency responder vital statistics (obtained automatically by physiological sensors located in garments), ambient temperature and environmental hazard data, remaining air supply, and other data to a central dispatch center where those data could be monitored. One participant added that while such solutions may be seen as being costly, the benefits from reduced responder injuries and deaths were likely to outweigh the costs.

HUMAN FACTORS

Another interesting issue that emerged from the community discussions as having important implications for personal protection of emergency responders is how personal discretion and decisionmaking can be critical determinants of PPT effectiveness. This section discusses some of these important human factors, including knowledge management (effectively utilizing available information), safety practices and enforcement, and the influence of service traditions and organizational culture.

Knowledge Management

As discussed in the previous sections, more advanced technologies for hazard information gathering are becoming available in the emergency response community. Yet, RAND's discussions with members of the community suggested that very few services have yet to take full advantage of this trend. Several reasons for this were identified.

As noted earlier, one impediment to acquiring such technologies is the initial cost, especially for large departments. Another limitation to utilizing more-advanced technologies is the ongoing cost and complexity of maintaining such systems. Several participants noted that an information management system is only as good as the information in it, but updating databases is an expensive task. Several fire departments with which we met had mobile computer terminals, but some departments rarely used them because little useful information was available on the system.

Personal views on the value of the information that dispatchers relay varied considerably. Inadequate training of medical dispatchers, said one participant, resulted in insufficient information on site conditions and risks being obtained from callers and subsequently relayed to responders. Another noted that the value of the information depended in part on whether the dispatch center was operated by the responding service.[4] Similarly, the challenge for pre-planning efforts was assuring that the information in the plans was regularly updated and maintained. One recommended solution was that the maintenance of hazard and building information systems needs to extend beyond the emergency response departments so that the information can be input by a variety of municipal services, such as building code enforcement, water, power, and transportation services.

> Forty rads, is that bad? I need to know if it's good or bad. If we were nuclear physicists, we wouldn't be cops.
>
> —*Law enforcement representative*

The discussion of information systems also focused on training of personnel to use them. A small-city fire service evaluated the performance of a computerized personnel accountability system that also maintained data on the health condition of responders. The system was seen as performing more functions than responders could realistically use in the field. The fire service opted to adhere to the "keep it simple" strategy and continued using accountability tags mounted on the back of firefighters' helmets. A fire service in a mid-size city reported having approximately 100 hazard detection devices in service ranging from more-basic technologies to sophisticated equipment. "We can deal with just about anything out there," said a representative, but he added that some of the equipment is not used because there are not enough line personnel trained to use it. One big-city fire department had acquired a number of $75,000 chemical analyzers to be used in the field. However, a representative of that department observed, "The complexity of the equipment is really overloading the normal firefighter. We are at the point where we really need to be able to hire lab technicians." However, the city's hiring and pay practices made it difficult for the department to acquire, train, and retain such skilled personnel.

> To me, simple is better. Firefighters have a tendency to shy away from things that are complicated.
>
> —*Fire service leader*

Many participants argued that environmental monitoring technologies should be simple to operate and understand: "No interpretation, no choices, just on or off," said one participant in referring to how a device should operate. "Ease of

---

[4]Dispatch services are often consolidated in such a way that a single department or third-party service may dispatch for the police, fire, and EMS departments in a city or even in multiple cities.

use is a big deal," said another participant. A law enforcement representative said he wanted hazard monitoring equipment to supply real-world answers, not numbers, that provide guidance on what to do. Other participants (including those with hazmat expertise) expressed doubts that such simplified ("firefighter friendly") approaches were realistic. For example, it was noted that knowledge of the concentration of a hazardous agent is by itself of limited use, and translating the reading into actionable information, such as "don SCBA" or "high explosion hazard," is a complex process that is fraught with uncertainties and requires specialized training.

Safety Practices and Enforcement

> You have to have discipline. There must be procedures established beforehand. Those procedures must be analyzed. The procedures must be implemented by everyone. We don't have that. We don't have the education. We don't have the training. We don't have command system. We don't investigate [injury] incidents. We have a lot of gaps.
>
> —*Fire service leader*

The RAND discussions tended to focus on high-visibility health and safety risks (such as thermal and chemical hazards) responders face at the scene of an event and advanced technologies to address those risks. Of particular concern was reducing the likelihood of acute injuries and death. However, discussions with several agencies pointed to chronic and day-to-day risks and occurrences that, although typically less acute than high-visibility risks, are often more prevalent. Such day-to-day risks also typically require more basic technology and practice solutions. A number of diverse issues were raised in this area, including the following:

- A person responsible for safety reported that shoulder injuries, followed by knee injuries, were a significant problem in his fire department: "If I could teach them to pull hose properly, I could dramatically reduce shoulder injuries."

- Another individual reported that a significant proportion of injuries in his fire department were incurred while firefighters were getting on and off apparatuses.

- Back and knee injuries are the largest sources of workers' compensation claims for small-city fire/emergency medical services, and those injuries were mostly attributable to moving and lifting patients.

Such injuries can take a toll on a responder's career. Injuries also were seen as imposing significant costs on an organization in terms of compensated sickness time, workers' compensation insurance and payouts, overtime costs associated with filling open shifts, and additional costs of training replacement personnel.

Several departments noted that they are actively trying to reduce workplace injuries for three reasons: (1) in an effort to cut lost work time, liability claims, and litigation; (2) as part of city and state safety programs; or (3) as a result of regulatory sanctions. Efforts being made to reduce slips, trips, and falls include improving station lighting, keeping station floors clean and dry, installing grab bars and nonskid surfaces on apparatuses, ensuring optimal fit of bunker pants, and selecting footwear with enhanced ankle support and wide heels.

Despite the growing desire to improve workplace safety and health, several agencies reported that they lacked the funds to properly monitor and analyze injury data and workplace practices or to implement programs to address the issue. One fire department leader acknowledged, "If we did a better job analyzing injuries, we would have a better understanding of what caused them."

> We put them in a competitive environment and they often end up driving over their ability. It is probably the one aspect of police service that we want to regulate the most.
>
> —*Law enforcement representative*

Motor vehicle accidents are a major cause of death and injury for all emergency responders. As noted in Chapter Two, 40 percent of police officer deaths, 18 percent of firefighter deaths, and approximately 50 percent of emergency medical service responder deaths involve motor vehicles. Motor vehicle accidents caused by emergency responders were of concern to several participants in the RAND discussions. Several participants noted that young police officers are tempted to drive too aggressively in a "lights and siren" situation, resulting in a significant share of officer injuries occurring in traffic accidents. Injuries (sometimes also to innocent bystanders) and issues concerning judgments and liabilities had motivated one department to pay more attention to driving behavior and implement a "very restrictive" driving policy. Other police departments also mentioned their driving and pursuit policies.

Traffic accidents were also cited as an issue for firefighters and EMS personnel, especially volunteers. In EMS, one participant asserted, the challenge in data analysis is in discerning the benefits versus the costs of driving at excessive speed. Driving practices used by emergency responders in "lights and sirens" conditions, such as driving at high speed, driving against oncoming traffic, and running red lights makes a positive difference in medical outcomes for only a small fraction of transported patients, he argued, but adds significantly to risks to emergency responders and bystanders. In all cases, better driver training and enforcement were seen as part of the solution.

> Accidents occur not because we don't have the equipment, but because we don't use the equipment.
>
> —*Fire service leader*

Enforcement of safe practices was cited by many participants as an area for improvement to prevent injuries of all types. One small-city fire department representative indicated that many injuries occur because firefighters on his force do not wear personal protective equipment (PPE) properly. Photographs of fire scenes, he added, were reviewed to identify problems in this area. A police representative noted that mobile computer terminals create a distraction inside the patrol car and their improper use may cause accidents. "Our policy says [officers] must pull over before they look at the screen or type. Do they? No." At major incidents, one participant said, not all firefighters in his department adhere to rules governing rest and rehabilitation: "If we follow our policy, then it works okay."

Rather than acquiring more or better protective gear, enforcement of safety rules and regulations was viewed by some participants as the top priority for improving responder health and safety. "Training and compliance are the biggest needs. We carry the necessary equipment," said one battalion chief. "You can carry all the stuff in the world, but you've got to wear it," argued another law enforcement representative. "They aren't very conscientious about wearing PPE," said a small-city fire service safety officer, during a discussion of exposures to pathogens. "It's a battle to get [firefighters] to wear seat belts," said another fire service representative. One fire department representative in a small suburb noted that his service responds to less than one major structural fire per year. This has resulted in laxness in using structural PPE.

Responder Fitness and Wellness

The need to improve protective gear was a principal concern of rank-and-file firefighters, patrol officers, and emergency medical service responders. However, improving protective gear and practices to reduce responder injuries and death will address only part of the risk equation. As discussed in Chapter Two, physical stress, overexertion, and cardiovascular incidents are the leading causes of firefighter death and injury and are significant causes of injuries and death among police and EMS responders as well. Many discussion participants remarked that strengthening the person and increasing his or her resilience and resistance to stress, fatigue, and injury are key to responder health and safety.

> We are much more fit than we used to be 15 or 20 years ago. It's a whole new generation of people.
>
> —*Law enforcement representative*

Many community representatives in leadership positions talked about the impact of fitness, diet, psychological stress, and other personal factors (described broadly as "wellness") on responder performance, resilience, and risk of injury or death in the line of duty. Those representatives also acknowledged that responder fitness and wellness, like safety and enforcement of safe practices, are major organizational challenges. Many participants pointed to their investment in physical fitness facilities and programs as one means to reduce injuries.[5] At the same time, many participants noted that they had mixed results in trying to achieve fitness and wellness objectives.

> A chief would rather tell the widow that the firefighter died battling a fire, than tell the firefighter that he can't do it anymore [because he is not fit].
>
> —*Fire service representative*

In many jurisdictions, fitness is a very sensitive subject. Commanders are hesitant to exclude members of their team from their line jobs, especially because alternative nonoperational support positions are relatively scarce and are often considered to be less prestigious than line positions. Requiring personnel to take and pass a medical exam is also a sensitive issue. Unions are concerned about maintaining fairness and protecting their members' employment status. "I've never seen an officer kicked off the department [because he or she] couldn't pass the physical fitness standard," observed one police department leader. Some participants attributed success in promoting wellness programs, or a lack thereof, to the culture or leadership of an organization, or to groups of individuals within it. Some pointed to regional and generational factors. For example, fire departments in the Northeast have a reputation of being more tradition-bound and resistant to introducing physical fitness programs than departments in other parts of the country.

In recognizing the need to address physical fitness in the fire service, and the sensitivity of the issue, the International Association of Fire Fighters and the

---

[5]Interestingly, one large fire department reported a concern with injuries sustained during fitness workouts, and it banned basketball playing during work hours.

International Association of Fire Chiefs have implemented some nonpunitive, voluntary fitness initiatives (Runnels, 2003). These initiatives include an overall wellness program addressing physical and mental fitness, a candidate physical ability test to help evaluate the fitness of potential firefighters, and a peer fitness trainer certification program to establish a basis for effectively training firefighters in physical fitness. Despite these efforts, most firefighters do not participate in a fitness program: A recent survey indicates that 792,000 firefighters (approximately 75 percent) serve in departments with no program to maintain basic firefighter fitness and health (U.S. Fire Administration and National Fire Protection Association, 2002). Importantly, the wellness program and candidate test initiatives will soon include data-tracking components that will help evaluate the effectiveness of the programs.

Tradition and Organizational Culture

During their discussions with RAND, participants frequently referred to the importance of tradition and culture in the emergency response community, which impacts the selection and use of personal protective technologies. Accordingly, improving the health and safety of emergency responders should not be seen as solely a technology issue but one that must also address tradition and culture.

> You know what they say about the fire service: "200 years of tradition unimpeded by progress."
>
> For some reason firefighters like clean apparatuses and dirty gear. It's a source of pride.
>
> —*Fire service representatives*

A prime example of tradition is the wearing of leather helmets with large brims, which are preferred by many firefighters because they hearken back to the early days of professional firefighting, and because wearers are readily identified as firefighters. Many fire departments reported an increased interest in such helmets in the aftermath of the September 11 attacks, despite the view of many responders that those helmets are heavier than more-modern designs. In addition, the larger brims and higher profiles of the traditional helmet design can restrict movement in some situations, prompting users to remove their helmets in order to do their jobs.[6]

---

[6]One firefighter observed, "What is the first thing you do when you need to enter a vehicle at a collision scene? Take off your helmet."

Dirty, worn, and burned gear was regarded by many respondents as a sign of experience and a "badge of courage." Said one EMS manager, "I've got a helmet that's pretty beat up, but I like it." However, a fire department official responsible for PPT acquisition noted that such preferences often result in personnel accepting the fact that they have a heightened exposure to hazards.

> People still use turnouts for everything. [A new type of certified protective clothing] is ideal for nonfire use, but firefighters don't use it. They cling to bunker gear like a security blanket.
>
> *—PPT supplier*

Several fire service participants argued that tradition can impact the diffusion of new technologies and practices. Many services have long-term attachments to specific equipment and relationships with equipment suppliers that they are very reluctant to abandon. "I like yellow. I've always had yellow," said a fire department leader about his bunker gear. When conceiving new technologies and procedures, developers need to keep in mind how they will be received by the rank and file. "It's tradition. That's what impedes the progress of PPE," said one participant from the fire service. "If it doesn't look the same, it won't be accepted."

Tradition and culture extend to safety practices and the perception and assumption of risk. "Safety is pretty subjective," said one fire service representative. Several participants drew a comparison between first responders' urge to render assistance and the need to take precautionary measures. "It's very hard to arrive on the scene where somebody needs assistance and contain yourself," said a police commander. "[If] you see victims, you don't think PPE," said another law enforcement representative.

Similarly, enforcing compliance with PPT use in emergency response was seen as being governed by organizational culture, given the fraternal and often voluntary nature of the profession. In a volunteer department, there are no financial penalties for violating the rules. Said one participant, "What are you going to do in a volunteer association? Take away their birthday?" Others raised the concern that the high-stakes culture made emergency response more risky than it needs to be. Speaking of firefighters and EMS responders in his agency, one participant stated, "They aren't very conscientious about wearing PPE. Firefighting is still regarded as a sport."

Some participants raised questions about the offensive "rush in" approach to structural firefighting and law enforcement that is dominant in the United States. In contrast, hazmat and bomb squad representatives emphasized their more defensive and cautious approach to situations, their recognition of the importance of hazard assessment in advance of taking action, and their rigorous adherence to safety protocols.

Interestingly, many participants raised the specter of moral hazard—that is, improvements in personal protective technologies and their use provide a form of insurance that may increase risk-taking. Participants worried that the excellent thermal protection afforded by their protective ensemble has enabled firefighters to overextend themselves in high-risk environments. "A real concern of mine is a false [sense of] security, that [firefighters] feel like they are bulletproof," stated a leader of a small fire department. Despite the fact that escape hoods are designed to be used only for very short durations during an escape, law enforcement representatives suspected that the hoods would be used by police officers for operational purposes rather than for immediately exiting a hazard zone.

Organizational culture and tradition, and therefore organizational practices, vary widely by service and region, it was noted throughout the discussions. One factor affecting the willingness to accept new PPT is the age of the responder: Younger personnel, it was frequently argued, tend to be more receptive to new equipment and workplace practices, such as wearing seat belts or ballistic vests. One participant noted that his department was paying more attention to cleaning bunker gear because the public expected firefighters to have a tidy appearance, driven in part by their involvement in various "public service" functions in addition to emergency response. In general, discussion participants broadly attributed variations in PPT practices to organizational culture, leadership, procedures, training, or local experience, such as notable responder injuries or deaths. Given such uncertainty concerning proper practices to follow, one participant recommended that research be funded to identify and document (i.e., benchmark) best safety practices across the emergency response community—a practice that is common in industry.

Chapter Eight

PROCUREMENT AND LOGISTICS

A surge in federal funding combined with a perceived heightened vulnerability at the local level since September 11, 2001, is pushing personal protective technology acquisitions into new territory for many communities. Decisions regarding how personal protection technologies are identified, acquired, and used in the field vary significantly among agencies, many study participants noted. Numerous issues and concerns were raised that have implications for PPT research and development needs. This chapter addresses issues surrounding the procurement; certification; and storage, transportation, and maintenance (collectively referred to here as "logistics") of personal protective technologies.

THE ACQUISITIONS PROCESS

Our discussions with participants uncovered some key areas of the acquisitions process that are in need of improvement: risk assessment and identification and evaluation of personal protective technology options.

Risk Assessment

Police don't have enough chemical protection. We don't even know what protection we need. We need information.

—*Law enforcement representative*

PPE is ordered according to tradition and personal preferences and is not linked to performance standards. Risk assessments are not done as part of the procurement process.

—*Fire service supplier*

In their discussions with RAND, participants indicated that few emergency response agencies have the resources or capabilities to conduct formal risk assessments to guide decisionmaking for PPT identification, assessment, acquisition, and deployment. In the fire service, for example, several participants claimed that protective technology acquisitions were based largely on tradition, style preferences, and inertia. Missing from the process, they noted, was an assessment of the risks that firefighters face, the protection they need, and the performance requirements for that protection. In the realm of terrorism response, the perceived threats driving PPT acquisitions are poorly characterized, and the protocols and training for PPT use are often not well developed or implemented.

An example of the uncertainties in risk assessment is the push to acquire chemical protection for terrorism response without having good models of the threat or plans for how the protection would be used. According to one participant from a major metropolitan police force, terrorism is "the biggest issue in law enforcement today. We are trying to determine what that means for us."

Personal Protective Technology Identification and Evaluation

Numerous participants described identifying and evaluating protective technologies as areas of the acquisition process that needed improvement. Strong loyalties and tradition, especially within the fire service, motivated agencies to stick with the suppliers they had historically used. The municipal acquisitions process can exacerbate this situation by requiring additional justification for purchasing new or different technologies. Systematic methods for evaluating technology options are not well established.

> People are buying stuff because they think, "That's the way it's supposed to look."
>
> —*Law enforcement representative*

When asked how they assess PPT performance before making an acquisition, most participants relied on information provided by suppliers or vendors. Yet, several participants were concerned that vendors' product claims were not properly justified. One fire department leader questioned the appropriateness of marketing literature that shows firefighters standing amidst flames: "In my opinion," he said, "it's gone too far."

As a result, vendor information often is backed up by personal references. This informal system of performance verification is maintained via e-mail, telephone, or encounters at conferences and meetings. Two participants noted, however, that such informal information was not always reliable because many individuals in the emergency response community are wedded to certain manufacturers and practices and are therefore biased in their opinions. In addition, while "most departments are open with their information," a fire service representative said, some are reluctant to report bad experiences. "It's kept on the hush-hush," he said. "People don't want to point the finger." One participant said that when he started digging into SCBA recall data, he discovered problems that were not reported by the professional media. "Nobody's got a perfect system out there, regardless of what they say."

Most agencies reported having review committees and conducting in-house studies and pilot trials of protective equipment, principally to check ergonomics, comfort, and the general receptiveness of the rank and file. Except for a few very large departments, most municipal agencies do not have bench-test facilities, trained analytic personnel, or funding to carry out rigorous performance assessments. Larger agencies might also hire a consultant to assist them with their PPT identification and evaluation process. Uncertainty about PPT performance was cited by two participants as a significant impediment to the diffusion of new technologies. "They want research to the nth degree," said one participant of municipal purchasing managers' need for thorough justification of a change in equipment.

> [When acquiring new technology] it's very typical for a department to get five different models and try them all to decide which is best. They reinvent the wheel—why should every department have to go through this?
>
> —*Fire service representative*

One issue further complicating the acquisitions process is that the effectiveness of much PPT equipment remains uncertain. Among the most critical examples of PPT equipment with uncertain effectiveness is environmental hazard monitoring technologies. In talking about the surge in spending on such equipment post–September 11, one participant who is knowledgeable about fire service technology development and standards said flatly, "These guys are going out and buying stuff, and they are buying junk." During the anthrax episodes in fall 2001, officials in one state banned the use of "smart-ticket" technology for testing suspected anthrax samples because of reliability problems, calling instead for all samples to be sent to a state laboratory for traditional culture tests.

Even technologies that are subject to rigorous standards, such as firefighter turnouts, are not regulated or monitored beyond the initial testing phase. If a defect or other problem is discovered after the technology is fielded, no regulatory or oversight agent is charged with notification or recall authority. Consequently, information about inferior performance or even catastrophic failure has not always been shared with the community. Similarly, no organization reviews PPT manufacturers' advertising and performance claims in the way that the U.S. Food and Drug Administration reviews claims about the efficacy of drugs and medical devices.

There's lots of junk on the market. We want a *Consumer Reports* system to rate PPT.

—Law enforcement representative

In response to these problems, many participants strongly advocated implementing objective, third-party assessments—akin to assessments by the Consumers Union and its publication *Consumer Reports*—to help guide them in their PPT evaluations and decisionmaking. To this point, the National Institute of Justice (NIJ), in conjunction with the National Institute of Standards and Technology, the InterAgency Board for Equipment Standardization and InterOperability, the U.S. Army Soldier and Biological Chemical Command, and the Technical Support Working Group, has compiled a resource guide for comparing PPT used for chemical warfare agents, toxic industrial materials, and biological agents (National Institute of Justice, 2002). This guide, one of a series of NIJ resource guides on technologies for emergency responders, offers data on duration of protection, dexterity and mobility, launderability, and use and reuse characteristics for commercially available equipment.

The NIJ guide largely comprises a detailed market survey, but it represents an important first step toward performance evaluation, given that many RAND participants indicated a need for additional information to help guide PPT acquisitions. Over the longer term (several years), the National Institute of Justice plans to subject selected protective equipment for law enforcement to laboratory testing and evaluation according to protocols to be established by the National Institute for Standards and Technology. This measure should greatly simplify the evaluation process and help address responders' questions about suppliers' claims concerning PPT performance.[1]

STANDARDS AND CERTIFICATION

The NFPA, NIJ, and NIOSH standards and certification play an important role in guiding PPT acquisitions, particularly in fire protection. Despite the emphasis on high-quality standards for firefighting and other protective garments, existing standards were seen as inadequate in the areas of ergonomics and sizing. Currently, turnout clothing is designed, tested, and certified according to standards that call for testing only swatches of fabric or testing garments in a static standing position. However, firefighting entails exposures to the entire body

---

[1]Performance evaluation of commercially available PPT is within NPPTL's mandate, although such evaluations are restricted to technologies for which NPPTL does not develop standards or certify equipment.

and is extremely dynamic. It involves tasks such as pulling down ceilings, wielding an axe, crouching, and crawling, one participant pointed out. Addressing such concerns will require improved clothing design. One participant suggested adopting designs from sportswear manufacturers that would improve the functionality of compression and expansion areas of a garment. In addition, new testing and certification procedures will be required to adequately address problems with exposure and ergonomics.

Consistency of sizing is also a problem, especially for women, who are being sized for garments as if they are small men. "We throw a sack of potatoes on someone. If you had a more custom fit suit, it's likely to be worn more," one representative noted. An inadequate range of glove and boot sizes and designs was a particular area of concern for several participants. "Picking the right kind of boot for 1,000 people and then stocking enough [of them] is a logistical nightmare," a participant said. One fire service representative went so far as to suggest the adoption of a national standard firefighting uniform, as has been done by the U.S. military and U.S. Forest Service and by national fire services in the United Kingdom, France, and Japan. Such a standard, it was argued, would reduce the vagaries of the acquisitions process and serve to push down prices. "Every city has to design its own gear," said one participant. "Everybody and their mother has a different interpretation of what's appropriate."

Compared with the fire service, the emergency medical and law enforcement services have less-well-developed PPT standards and certification programs. While NIJ compiles information for a number of law enforcement technologies, it gives limited attention to biological, chemical, or respiratory protection. The absence of national bodies (analogous to NFPA) focused primarily on personal protection guidelines and standards, or even a common view of what is appropriate in these areas, contributes to the great variability in strategies for levels of personal protective equipment used. "It's so fragmented," said one EMS community member of the situation. Given the lack of guidance from within the profession, decisionmakers in those services turn to outside organizations, such as the North Atlantic Treaty Organization (NATO), the Occupational Safety and Health Administration (OSHA), private industries for industrial (e.g., hospital-based) safety models, or NFPA, for guidance.

Standards and certification also remain lacking for major classes of fire protection such as environmental hazard monitors and other electronic devices. NFPA has recognized this deficiency as an important issue and has recently organized a committee to address it.[2]

---

[2]The NFPA Committee on Electronic Safety Equipment for Fire and Emergency Services is responsible for documents on the design, performance, testing, certification, selection, care, and maintenance of electronic safety equipment used by fire and emergency services personnel.

LOGISTICS

Emergency responder organizations are acquiring more PPT—increasing the supplies of gear they already have on hand and acquiring new technologies—often with the assistance of federal and state funding. This creates a new set of questions: How do you store and maintain all this new equipment? How do you transport it? How do you outfit responders so they can operate with all this gear?

Storage, Transportation, and Outfitting

> You have to gain weight to get everything on your belt.
>
> We are getting more and more equipment every day. Sometimes too much.
>
> —*Law enforcement representatives*

Many participants pointed out that personal protective equipment must be readily available when it is needed, otherwise emergency responders are unlikely to use it. "If you don't put PPE on before you leave the barn, you won't use it," said one participant. Yet, as emergency responders have acquired more PPT, they have become increasingly loaded down with gear. Firefighters spoke of the desire for extra pockets, hooks, and belts to handle specialized gear. While enhanced protection was seen as being desirable, many fire and police representatives also raised concerns about their reduced effectiveness from being excessively encumbered. To this point, one department had substantially improved compliance in emergency medical response by issuing fanny packs containing an ensemble of protective gear.

Vehicles, too, are becoming increasingly crowded: Fire apparatuses are becoming increasingly full of equipment, EMS vehicles have very limited storage space, and police patrol cars typically carry cones, flares, first-aid equipment, and crowd-control gear. "We are getting to where we need a trailer on the back of the car," said one senior law enforcement representative. Police officers mentioned that gear stored in squad cars gets knocked around and damaged and exposed to the elements, and because patrol cars often are pooled, personal protective equipment frequently gets lost. One large metro police department that participated in the RAND discussions is issuing all patrol officers Level-C hazmat gear and air-purifying respirators. "How quickly will this stuff deteriorate in a trunk that is 140 to 160 degrees?" asked one officer. Several other departments voiced similar concerns. "Gloves, gowns, and masks are supposed to be in the [squad] car," said a representative of another major metro police force, but they often disappear. Respirator face pieces frequently get cracked or contaminated with food. One solution being used to reduce the storage space requirements and enhance the preservation of PPT is shrink-wrapping of gear.

> Where are you going to park all of this stuff?
>
> —*Fire service leader*

Many communities have purchased dedicated disaster response vehicles and trailers, and many have created supplemental caches, but these solutions raise questions about how rapidly the equipment will be fielded and who will have access to it. In the case of a serious chemical event, stockpiled equipment is of no good if it is not out in the field, argued an emergency planner. When an event happens, EMS personnel typically use only what is on their vehicles, said one participant. This was a key argument supporting why one EMS service was equipping each of its first-response trucks with a large duffel bag filled with PPT for use in case of a WMD-type scenario. "You need to have it right now. Space is a hot commodity, but we are putting up with it," said a representative of the service. In addition to acquiring equipment caches and vehicles, agencies need environmentally controlled garages and warehouses to house and maintain them—expensive capital expenditures for most municipalities. Moreover, facilities costs usually are not reimbursable under state and federal PPT assistance programs.

Maintenance and Reliability

As emergency services acquire more gear, their maintenance and reliability costs are increasing. Having more gear in the field means more gear will break or wear out over time. Many municipal services reported buying Level-B hazmat suits for first responders, but one group of firefighters questioned whether those suits would be serviceable after being used and then folded up and stored. A police department representative questioned the effectiveness of escape hoods and air bottles that had been stored in patrol cars for six or seven years, noting that many were overdue for testing. Another large police department listed equipment care as a real concern, noting that respiratory and chemical gear may sit in a trunk for days after being used. More-sophisticated gear, such as environmental monitoring equipment, also requires complex and expensive testing, calibration, and repair, capabilities that few departments have in house.

The cost of spare parts, such as replacement batteries to power the increasing number of electronic devices, was cited as being significant by both a fire department and a police department.[3] Even for small-cost items, a lack of authorized funding for spare parts and repair is causing problems. For example, one police representative noted that face shields lose screws from getting knocked around in patrol cars, but no funds are available to replace them. "We don't have money for that," he said. Advanced medications require periodic replacement. Along those lines, a big-city medical strike team was having problems managing its pharmaceuticals inventory, leading to questions about the efficacy of the stocks on hand. One person in the team noted, "As a first responder, I would like to know that the drugs are available."

The pace of PPT turnover is being accelerated because of management concerns about liability for equipment failures in the field and a new NFPA standard (1851) specifying procedures for PPT care, inspection, maintenance, and replacement. Training and retaining maintenance personnel can be very expensive. One participant noted that firefighters are capable of performing daily maintenance of respirators, but specialists are required to perform periodic equipment overhauls and advanced maintenance when, for example, there is a high-temperature exposure. A small-town police official said he wished his city had a "czar" to tell his department which safety equipment needed testing and maintenance.

Increasing emphasis on maintenance and reliability led several participants to call for simpler ways to inspect gear and more obvious ways to identify existing or imminent failures. An example commonly cited by numerous fire departments was the desire for easy inspection of the moisture barrier in turnouts, a concern driven by recent incidents of serious burns being caused by undetected decomposition of moisture barriers. Similarly, several manufacturers as well as fire and police departments felt that passive integrity monitors would be a valuable addition to protective equipment. Participants pointed out the utility of equipping aluminum ladders with temperature-sensitive tags that change color as conditions change to warn of potential heat damage. They felt that analogous systems would be very useful as alarms to warn the wearer of failure or expiration of components, such as turnouts, chemical-protective clothing, helmets, and respirator cartridges.

---

[3]A police official noted that replacement batteries for flashlights cost his department $78 apiece. A fire official complained that replacement batteries certified as intrinsically safe for the radios used in his department cost twice as much as the noncertified variety.

> We have a lot of exotic equipment that is used, and not used, that is not maintained adequately.
>
> —*Fire service representative*

Care, inspection, maintenance, repair, and replacement of PPT typically are classified as operations or overhead expenses, where they compete for funding with numerous other priorities. Explicit funding of such critical tasks usually is difficult for agencies to justify to elected officials and taxpayers. Moreover, many participants wondered whether financial resources will be available for restocking PPT several years down the road after the post-9/11 concern with homeland security recedes. One participant noted that the U.S. Drug Enforcement Agency had cut back on funding for personal protection equipment and training even while operations involving drug laboratories remained a serious problem. "It's killing us," the participant said. "There is no money for sustainability. If you don't build in sustainability, you are actually hurting us," said another responder.

Given these pressures, many participants noted that PPT replacement decisions seemed to be arbitrary and not necessarily reflective of the true performance capabilities of personal protection equipment. Several participants expressed their suspicion that manufacturer-recommended service intervals and shelf lives were shorter than necessary, reflecting an interest on the part of manufacturers to reduce liability and increase sales. One police department representative cited a case in which ostensibly the same respirator cartridge was assigned a shelf life for military use that was considerably longer than that for municipal use. Rescue ropes, according to NFPA guidelines, must be cut up and discarded after a single use or after five years in storage. A fire service representative noted that inspection requirements were motivating his department to discard bunker gear before the end of the gear's service life to avoid having to go through required inspection protocols. One very large police department reported doing periodic inspecting and testing of respirator cartridges and returning to service those batches that passed inspection. Most municipal agencies, however, do not have such capabilities or resources.

RISK-SPECIFIC VERSUS UNIVERSAL EQUIPMENT

> Every place they go, they wear structural firefighting gear. Is structural firefighting gear appropriate when a person goes out in a boat?
>
> —*Fire service representative*

As discussed earlier in this report, the workload that emergency responders carry and the risks they encounter are changing. The number of structural fire responses is decreasing while the number of medical calls is increasing. Responders are concerned about the increasing risk of communicable diseases. Counterterrorism protection is dictating many departments' priorities.

At the same time, the ability to assess and manage these risks is also improving as the quality of available information improves and preparation of rank-and-file emergency response workers also improves. As one technical expert noted, "There is more of an assessment function happening now compared to the past. People know more now about hazards and that there are different levels of hazards, so a fixed single standard is not reasonable." In response, many organizations as well as individual responders are seeking more-varied levels of protection and greater flexibility (a "menu-driven" approach) in choosing among personal protection technology options for discrete hazards. Tasks with the potential for greater use of risk-specific or "tailored" PPT that were mentioned in the discussions include fire attack, ventilation, overhaul, medical emergencies, automobile accidents, urban search and rescue and technical rescue, WMD events and public decontamination, and wildland fires.

The standard-issue protection in the fire service is "universal" protection, or a single ensemble designed to protect against all anticipated hazards. Consequently, a firefighter's protection options are maximum protection or no protection. Nonetheless, many responders said they modify their equipment and ensembles for some tasks because they felt that the maximal protection offered by their standard-issue gear was too burdensome. Because of these limited options, many responders reported being underprotected in some situations. Commenting on this point, one firefighter stated flatly that "PPE protects against death, but not disability." A common example of underprotection that was cited is firefighters deciding to eschew SCBAs during the overhaul phase of a structural fire response. Municipal fire services (especially in the West and Southwest) are engaged in an increasing number of wildland fire events of increasing intensity as more homes and businesses are built "in the trees."[4] One

---

[4]The intensity of wildland fires has increased due to the build-up of fuel caused by decades of fire suppression and, more recently, public resistance to controlled burning and thinning.

mid-size municipal department reported taking a hybrid approach to its wild-land fire ensemble: mixing National Forest Service–style wildland shirts and goggles with standard structural firefighting bunker pants. Because of the high cost of equipping the entire fire department with a set of leather wildland boots, firefighters were given the option of buying and using their own. Tailoring and modifying personal protection equipment and practices are especially preva-lent in the emergency medical and special operations fields, where PPT stan-dards are less developed and greater emphasis is placed on individual respon-der decisionmaking.

Moving away from uniform, maximal PPT was recommended by many partici-pants. One fire service leader recommended three clothing ensembles and a three-tiered system of respiratory protection:

- Hot Zone: SCBA

- Warm Zone: Powered APR with face piece or hood (no fit test required) to provide protection from extended exposures

- Support Zone: APR (fit testing required).

In addition to having tailored gear, one fire service representative recom-mended having interchangeable ensemble components. It would be better, the representative said, "if you could layer or mix and match" components. To this point, one manufacturer recommended a modular protection approach that was being developed in London. The proposed system starts with a station uni-form and, by adding components, builds to discreet protection levels that are appropriate for emergency medical response, technical rescue, or structural firefighting.

Many individuals who participated in the RAND discussions expressly rejected the desire for risk-specific PPT in favor of universal PPT. Universal PPT, cur-rently the standard practice in the fire service, assures a uniform and high level of protection. Proponents of universal PPT also warned that more options for protection translated to more opportunities for mistakes. A single option re-duces the risk that a responder would not be familiar with its use, and it relieves responders from the need to make complex PPT decisions in what are often stressful and time-constrained environments.

Another argument proffered in opposition to tailored PPT is that hazard infor-mation often is not specific enough to select the appropriate PPT. "If you don't have information, you don't have the option of what PPE to use," said one par-ticipant. Further, responders often do not have a specific task assignment until they arrive at the scene, at which point, it was argued, there is no time to select and don specific PPT. Finally, increasing the number of PPT options will result in an increase in the amount of gear that a department must acquire, store, maintain, transport, and provide training on, which was seen as adding to an already heavy logistic burden on departments. For many departments, the decision to issue only one set of gear is strictly determined by funding constraints. However, despite opposing a tailored approach, proponents of universal PPT routinely confessed to a certain ambivalence over the situation. As expressed by one representative from a large fire department, "The problem is that there is not enough information to be sure [that alternate protection is appropriate]. Still, the present options are too limited."

MUTUAL AID AND INTEROPERABILITY

The recent terrorist attacks on the United States severely taxed the capabilities of local emergency response organizations. Responders present at those events noted the importance of mutual aid and raised the need for greater PPT interoperability and standardization—especially in the case of respirators (Jackson et al., 2002). Planning for mutual aid and pooling equipment are ways that local agencies can share the preparedness burdens, complement each others' capabilities, and lower their equipment and logistics costs. Interoperability and standardization would facilitate the sharing of equipment in the field and would facilitate the use of supplementary equipment delivered to a major event site (e.g., delivered from federal caches) by assuring compatibility with existing gear and requiring less training and fitting in the field.

> The last thing that people are standardizing is personal gear.
>
> —*Fire service leader*

Despite the potential advantages of interoperable protective equipment, the organizations with which we spoke rarely coordinate their acquisitions with other services within their community or with neighboring jurisdictions to facilitate equipment sharing in a mutual-aid situation. With neighboring departments making acquisitions independent of one another, little attention is paid to their complementary strengths, sometimes resulting in unnecessary redundancy in capabilities. The local police, fire, and EMS agencies "are all buying the same stuff," said one emergency management specialist. Meanwhile, agencies may suffer critical gaps elsewhere, noted an EMS representative.

This situation does not appear to reflect any purposeful efforts to avoid interoperability. On the contrary, when asked, most departments acknowledged that equipment compatibility and interoperability would be beneficial. Rather, the lack of PPT coordination in the emergency responder community stems from a number of formidable impediments. Major barriers to PPT coordination and interoperability that were cited in the discussions include the following:

- Communities' PPT purchasing cycles rarely coincide with each other and "use-it-or-lose-it" funding mechanisms deter long-term planning and coordination.

- Neighboring communities often have different purchasing power—for example, metropolitan centers may have greater buying power than small satellite communities.

- Changing of technologies often entails a major initial expense because new personal protective equipment, as well as ancillary support equipment and services, ideally must be changed throughout the entire department at the same time.

- Agencies' well-established vendor relationships and traditions favor certain PPT practices or manufacturers that often are incompatible with the practices or equipment of neighboring organizations.

- Logistical and bureaucratic questions, such as who pays, who receives, and who stores equipment, thwart cooperation.

- There is reluctance on the part of emergency response agencies to rely on an outside agency for particular capabilities.

These barriers to PPT coordination may be compounded by the small proportion of large mutual-aid events within the spectrum of emergency responder activities. Consequently, priorities for protecting emergency responders are focused on more-common events and protection needs.

A few exceptions to this situation were mentioned in the discussions. One municipal EMS service reported purchasing the same respirators as the local fire service: Many of the EMS personnel had been trained on SCBAs in the fire service before coming to the EMS service, and the EMS service relied on the fire service to manage its respirator maintenance needs. A fire/medical service covering a large territory reported sharing technical rescue and hazmat gear with neighboring jurisdictions "so that when you get on the scene there are no surprises." Fire departments in three neighboring cities of comparable size and economic means share responsibility for hazmat, light rescue and air utility, and USAR. Despite this cooperative effort, the agencies do not coordinate their equipment acquisitions except for those related to communications.

In general, however, coordination to enable mutual aid did not include PPT. As one participant explained, "In terms of emergency management, yes; in terms of training, yes; but in terms of PPT, we're not there." Several agencies did report using the same vendors as other jurisdictions in their area, but this situation occurred only because state regulations enabled them to obtain favorable prices by purchasing under the same contract.

In addition to the obstacles at the user end, several impediments to interoperability and standardization exist on the PPT supply side. First, manufacturers drive much of the PPT research and development in the emergency response field, and they have a strong financial interest in proprietary designs as a means of differentiating themselves in a crowded market and earning a return on their investment. This is particularly true given that the strict certification standards leave little room for various manufacturers to distinguish their products. When a department has made a substantial investment in a piece of proprietary technology that is functionally incompatible with competing options, it becomes difficult for that department to choose options from other suppliers, thus making that department less able to coordinate PPT acquisitions and use with neighboring jurisdictions. Many suppliers also are reluctant to promote interoperability because they wish to avoid any liability for systems over which they do not have complete design control. As one participant summed up, "The fire service is very dependent on manufacturers and manufacturers' interpretation of what is important."

Chapter Nine

PUTTING COMMUNITY VIEWS TO WORK

In the preceding chapters, we presented emergency responders' views of the risks they face in the line of duty and their most critical personal protection needs. As Chapter Two illustrated, the emergency response community is extremely diverse in the size and structure of its organizations, the populations it serves, the tasks it undertakes, and the hazards it encounters. Not surprisingly, the subsequent chapters brought to light important similarities and differences in emergency responders' views of risks and personal protection needs.

In this concluding chapter, we gather together these perspectives into a number of findings across broad issue areas, which collectively may be considered an initial step in developing a personal protection agenda for the emergency responder community. First, we offer observations on improving both equipment and practices, which together are defined as *personal protection technologies.* Most of these observations were originally put forth as recommendations, summarized in Table 9.1, from individuals who participated in the RAND discussions. Several observations were extrapolated from participants' comments. Next, we turn to several broader policy issues raised by the community discussions that are salient to the personal protection of emergency responders and that warrant further research, analysis, and discussion.

COMMUNITY PRIORITIES

Reducing Physical Stress and Improving Comfort

Personal protective equipment often is heavy and burdensome to the wearer and can cause physical stress and overexertion. Physical stress and overexertion are the top causes of injuries and deaths among firefighters, accounting for more than one-quarter of all injuries and almost one-half of deaths in the late 1990s (as was shown in Figure 2.4 in Chapter Two). Physical stress and heat dissipation were also major concerns for police officers wearing ballistic vests and hazmat personnel wearing chemical-protective clothing.

Table 9.1

Personal Protection Priorities and Recommendations Raised by the Emergency Responder Community

Personal Protection Priorities	Specific Recommendations
Reduce physical stress and improve comfort	• Improve garment breathability • Reduce equipment weight • Ensure consistent and appropriate sizing of components • Enhance ergonomic characteristics
Improve communications	• Make radio systems interoperable • Improve communications capabilities with SCBA • Improve radio design to allow hands-free use and use with gloves
Upgrade communicable disease protection	• Increase protective equipment options for EMS personnel and police
Develop practical respiratory and chemical protection equipment and guidelines for first responders	• Improve the chemical and biological protection of garments and respirators • Design protective equipment such that it minimizes interference with responder activities • Require more chemical/biological hazard training
Improve PPT standby performance	• Develop integrity monitoring and service-life monitoring technologies • Enhance compactness and portability of protective equipment • Address logistical complications • Reduce protective equipment maintenance complexity and cost
Expand training and education	• Require more training on sophisticated protective equipment • Reduce complexity of new equipment
Benchmark best safety practices	• Study and benchmark safety practices, particularly for EMS and police • Study and benchmark PPT enforcement practices

Reducing physical exertion and stress is becoming an increasingly critical concern of the responder community. Several discussion participants noted that while the frequency of structural fires is decreasing, the intensity of fire events is increasing, and the hazards found across the range of emergencies that responders are expected to face are becoming more complex. The recent terrorist attacks in the United States support this view: Responder activity stretched into days and weeks after the attacks on the Pentagon and World Trade Center. With the anthrax attacks, emergency responders all over the country were called out to a large number of "white powder events," which had their own associated demands and personal protection requirements. Overheating, foot blisters, "mask face," and neck strains from heavy helmets were among the many chronic ailments that diminished productivity and led many responders at the scenes to forego any protection at all (Jackson et al., 2002).

Keeping emergency responders physically fresh and unencumbered while providing adequate protection clearly is a challenge from a technology standpoint. With current personal protective technologies, increased thermal protection and barriers to toxic chemicals and biological agents (i.e., increased encapsulation) generally result in increased heat and moisture retention and discomfort. At the same time, emergency responders appear to be pleased with the primary protective functionality of much of their equipment. If anything, emergency responders raised concerns about *excessive* protection.

Rather than maximizing protection, the feeling of the emergency response community was that reducing physical stress on responders and increasing their comfort should now be made a top priority. Recommendations made by study participants to address the problem of excess physical stress included developing lighter-weight, more-flexible structural firefighting garments and ballistic vests and improving the heat dissipation and vapor transmission capability of those garments through better materials and construction. Other options included reducing and redistributing the weight of SCBA bottles and other components.

Beyond the personal protection gear itself, important procedural measures that were noted include keeping firefighters well hydrated and rested, and carefully monitoring their work cycles and physiological condition. To this end, responders suggested developing ways to monitor real-time health status—such as body temperature, heart rate, and respiration—during a response. Such information could powerfully inform decisionmakers at the command level about when to rest or rotate responders to avoid serious injury or death. Improved accountability systems that make use of advanced sensors, GPS, and communication technologies were seen as being critical to remotely tracking individual responders' location, activities, and condition. Other participants recommended a low-tech approach: better training and adherence to conventional rest-and-rehabilitation protocols.

Indeed, some participants argued that thermal protection afforded by the firefighter ensemble was too great (especially since the introduction of improved hoods) and had reached such a level that firefighters could no longer sense their environment through their equipment. One suggested solution was the use of "smart" coats with built-in temperature sensors and alarms. While these coats are currently available, they have not been well accepted because of reliability concerns. Several practice solutions also were put forth: more-realistic training opportunities involving real fires, a more cautious or defensive approach to structural firefighting, and closer adherence to already existing safety protocols.

Improving Communications

Communications problems were among the most consistently noted shortcomings by emergency responders, firefighters in particular. The communications issue is significant in regard to protecting emergency responders because communications help the community to gather and disseminate information and manage their response activities. Moreover, reliable communications will become more important in the future as more information and data become available and need to be shared. Two issues were cited on this point, one related to radio hardware and one related to communication system functionality and interoperability.

First, emergency responders mentioned several problems with handheld radios: They are difficult to use when wearing respiratory protection; they are difficult to operate while wearing gloves; they require the use of at least one hand, making them difficult to operate when the user's hands are engaged in other tasks (and making it difficult for the user to do anything else with his or her hands while operating the radios); and their controls are often inadvertently activated in the rugged environments in which they are used. Radio components and voice amplifiers integrated with the SCBA face piece were seen as offering only limited improvements in voice clarity and ease of use. These problems become more acute under the total-encapsulation environments required for hazmat and WMD response. Therefore, further development of integrated respirator-radio technologies with "hands-free" features and wireless connections was seen as being highly desirable.

Second, when acquiring communications systems, emergency response organizations must make trade-offs among selecting the most appropriate system in terms of price, performance, ease of integration, and other factors for their own jurisdictions. Being able to communicate with other agencies in mutual-aid scenarios is a lower priority. A common refrain heard in the discussions was that departments could not communicate with each other easily. Communication gaps exist among departments in the same city, among departments in neighboring jurisdictions, and among municipalities and state and federal agencies. There have been efforts, supported by the federal government, to make all emergency responders convert to 800-MHz trunked digital radio systems. These efforts have met with limited success because these systems

- are often seen as being prohibitively expensive (e.g., given the necessary up-front conversion costs)

- suffer performance problems (e.g., non–sight-line signal penetration)

- do not provide an "all-in-one" communications solution (e.g., agencies must still maintain parallel and backup systems).

Talk of a new 600-MHz communications option does not address the cost or integration concerns.

The general feeling of the emergency response community was that no solution to these problems is in sight. Communications systems are large, complex, and expensive, and the number of technology suppliers to the emergency response community is small. Moreover, the market for emergency response communications technologies is fragmented, and agencies' purchasing power is limited compared with industrial users. This limited purchasing power minimizes the community's ability to influence R&D and design decisions or influence pricing. To overcome performance and integration obstacles in the short term, municipalities that are still relying on analog systems must be motivated to convert to the 800-MHz technology *and* coordinate their efforts with other agencies. However, motivating municipalities to take these steps will require substantial financial incentives. To overcome these obstacles over the longer term, resources will have to be applied to a comprehensive emergency response communications R&D agenda that addresses interoperability, scalability, cost-effectiveness, reliability, and ease-of-use concerns.

Upgrading Communicable Disease Protection

According to the RAND discussion participants, the threat of communicable disease in routine emergency response is the top concern for emergency medical service responders and a primary concern among police as well. Of the range of hazards that was discussed in this study, the threat of communicable disease emerged as the one for which protection needs are greatest, in part because of the lack of viable protection options, particularly for the hands and face. For all the concern expressed by emergency responders, very little was mentioned in the way of recent innovations or recommendations. This lack of options for protection against communicable diseases may indicate that more fundamental research and development into fluid-borne pathogen protection is needed. It may also suggest that equipment is not the entire solution, and that a comprehensive approach including enhanced training and operational protocols is needed.

Developing Practical Respiratory and Chemical Protection Equipment and Guidelines for First Responders

Currently, a number of efforts are being made to improve WMD protection for emergency response. Most of these efforts center on high-tech solutions for emergency response specialty operations. A top priority for emergency responder departments is providing respiratory and other chemical and biological protection for first-responding police officers, firefighters, and emergency medical service personnel.

Historically, only specialized teams, such as those for hazmat response, have had training and access to advanced chemical and biological protection. These units are typically deployed only when circumstances indicate that special equipment and training are required. However, the majority of calls for assistance are handled by patrol officers, firefighters, and EMS personnel who have little or no chemical/biological protection or training. This situation has given rise to concern in agencies across the country that these first responders also need enhanced protection to deal with an increasing threat of terrorism and industrial accidents.

Despite the demand for better protection for first responders, participants pointed out that there is a critical lack of appropriate equipment, training, guidelines, and know-how to provide the level of protection required. Air-purifying respirators are items of principal interest in this area. While APRs are commonplace in industry—where their use is guided by rigorous environmental characterization and usage guidelines—little attention has been paid to the use of APRs in first-response scenarios. Yet, an emergency response environment is very different from the workplace environment in industry. In an emergency response scenario, chemical substances are likely to be unidentified, the users' physical surroundings are unpredictable, and there is likely very little time to don equipment or protective clothing. Moreover, the emergency response community repeatedly called for performance standards and certification procedures for chemical/biological protection that is appropriate for first-responder applications. Departments that had evaluated respiratory and chemical protection routinely expressed frustration over the fact that the available guidance is overly industry oriented. In response to these concerns, respiratory- and chemical-protective clothing standards for emergency responders are beginning to be developed.

Hazard detectors are a related technology of interest that first responders can readily use. Drawing on the experiences of industry and the medical profession, responders recommended the development and diffusion of passive "badge-type" detectors or sensors, which indicate when the user is exposed to chemical or radioactive contamination, and other rapid and easy-to-use environmental monitoring and risk-assessment technologies.

The conclusion reached from the discussions is that governmental guidance for respirators and chemical-protective clothing should be expanded to address the needs of first responders. This guidance should take into account not only the physical characteristics of chemical-protective clothing and equipment, such as fit, air filtration cartridge options, and ease of communications when wearing the equipment, but also operational considerations, such as how users are to be fitted and trained to use the equipment, when such equipment should be used, how to conduct operations when using the equipment, and when to evacuate a scene even when wearing such equipment.

Improving Personal Protective Technology Standby Performance

One need that emerged indirectly from the discussions is the need to improve the "standby performance" of protective gear. Improving emergency responder protection traditionally has emphasized the development of technologies and capabilities for use in emergency response activities. However, participants indicated that they desired greater PPT availability and readiness. These needs can be met by improving total service performance—that is, performance when equipment is being used and when it is not being used. Key priorities identified in this area include the following:

- Low cost

- Ruggedness and durability

- Compactness, light weight, and portability

- Ease of maintenance.

As mentioned earlier in this chapter, many participants said that the thermal protection currently provided by structural firefighting gear is adequate for most tasks for which it is designed. On the other hand, many participants pointed to the accelerated degradation of thermal protective garments when the garments are exposed to ultraviolet light. Police officers mentioned that PPT stored in the trunk of a squad car becomes damaged when evidence is placed in the trunk. They also noted that PPT becomes damaged when it is exposed to heat and moisture, and PPT is frequently lost. Thermal imagers and environmental monitors get dirty, banged up, and dropped.

As equipment becomes more complex, it also tends to become more susceptible to perturbations, and it tends to become more expensive to repair, creating undesirable trade-offs for emergency response organizations, especially smaller ones. Thus, one area for focused improvement is "ruggedizing" PPT to reduce the likelihood of damage, slow equipment degradation, and minimize maintenance requirements.

The emergence of new hazards and the introduction of new PPT raise concerns about PPT logistics—i.e., storage, transportability, and maintenance. The increasing amount of gear that is being deployed in emergency response is leading to storage and transportation problems in stations, on vehicles, and on individuals. Storage space is at a premium, and storerooms, vehicle trunks, and tool belts are running out of room. As we have shown in this report, if personal protective equipment is not readily available when it is needed, emergency responders are unlikely to use it. This situation suggests a priority need, at this point, to reduce the size and weight of PPT components, rather than increase protective equipment performance.

Concerns about escalating costs of PPT maintenance, which lead to trade-offs in the level of protection that is available to responders, point to the need for improvements in the areas of PPT maintenance and reliability. Many emergency responders expressed uncertainty about the status of their protective gear. They have limited real-time opportunities to test how the protective performance of PPE degrades over time due to normal wear, environmental assaults such as heat and moisture, and nonuse and storage. The PPE performance tests that are available often require destruction of the equipment, such as for inspection of bunker gear interior layers and respirator cartridges.

Respondents also lack independent validation of their judgment on whether or not to use certain equipment. Accordingly, they expressed interest in having ways to facilitate inspection and confirm integrity of equipment. For example, color-change indicators and other technologies that warn of impending failure or expired service life were viewed as worth pursuing by all members of the community. End-of-service-life indicator systems could simplify maintenance of a number of components, including turnouts, chemical-protective clothing, and respirator cartridges, by helping to assure that gear is not discarded too soon or held in service too long. Such systems would be valuable for equipment that is not used frequently, such as equipment for WMD events. Before such technologies will be used, however, they must be regarded as being highly reliable.

Expanding Training and Education

The subject of PPT and emergency response training and education came up repeatedly in many of the discussions. In sum, community representatives stressed that a greater amount of training and education must be made a part of any policy to improve the protection of emergency responders in the line of duty. Providing emergency responders with personal protection equipment without also providing proper training, it was argued, vitiates the equipment's effectiveness, and, at worst, is unethical.

Several challenges the community faces in this area were noted by participants:

First, skills maintenance is critical. PPT training, when provided, is often done at the front end when the equipment is first introduced and the responder receives his or her initial training, and refresher training is not done frequently enough. Community members pointed out that maintaining certain levels of training and expertise for a range of protective technologies and safe practices is difficult for most responder organizations, particularly those that are volunteer based. These issues are particularly troublesome with highly knowledge-intensive but infrequently used technologies (e.g., complex environmental monitoring devices) and in some decisionmaking circumstances (e.g., in industrial or WMD-type events).

Second, for the training to be most effective, it must include realistic "operational" or "situational" scenarios and simulations. For firefighters, actual fire scenarios are becoming increasingly important as the frequency of major fires decreases. For WMD and other special operations scenarios, exercises using a live agent, such as tear gas, were seen as being essential for testing decisionmaking and operations in high-stress environments. However, the inherent costs of such testing and the regulations governing it were said to be constraining opportunities for situational simulations. "To do rescue training the way you're supposed to is expensive," said an EMS representative.

Third, many participants spoke highly of the federal WMD terrorism training efforts provided since the late 1990s under the Nunn-Lugar-Domenici Act. Not only do such efforts facilitate PPT preparedness, and in many cases use live scenarios, they also offer valuable opportunities for community members to get to know each other. This indirect benefit facilitates potential future cooperation in mutual-aid situations and encourages cross-fertilization of ideas and know-how among local agencies. However, funding for such training activities, as was said earlier, is not large enough to maintain a sufficiently skilled cadre of personnel across the United States, given force turnover and staff rotation policies.

Fourth, for all types of training, local law enforcement agencies are at a particular disadvantage given their staffing policies. Adequate funds must be made available not only for the actual training activities, but also to cover the cost of trainees' time and the time of replacement personnel filling in for the trainees. As a result, agencies find it very difficult to train large groups of personnel and sustain competency across an entire force over time.

Finally, an issue raised by many participants throughout the emergency response community was the need for training and education to develop greater analytical capabilities in all quarters of the community. Protecting the health and safety of emergency responders traditionally has been approached through the development of protective equipment and standard operating procedures. Emergency responders have been trained to follow these procedures. Today, the threats responders face are more uncertain; the "human element," such as heavily armed assailants and terrorists aiming for maximum impact, is more aggressive than in the past; the potential risks to responders and communities are greater; and personal protection technologies are more complex. "We are starting to apply basic hazmat procedures to all calls," said one police department leader. As a result, emergency responders must rely on their own knowledge acquisition and problem-solving capabilities to a greater extent. As such, personal protection is becoming more dependent on enhanced threat awareness, detection, and identification; information sharing and analysis; and operational discretion and flexibility. All of these challenges call for fundamental changes in the intensity, frequency, and substance of personal protection training and education.

Benchmarking Best Safety Practices

Among all the emergency response services in the United States, the fire service is noted for having the most comprehensive personal protection equipment and practice standards. Nonetheless, in a number of areas, uniform practices have not been widely adopted in the fire service (even where standards are in place), and performance appears to vary greatly among fire departments. Police and emergency medical service representatives reported even greater variance in the employment of both personal protection equipment and practices. Those practices include:

- Workplace safety practices (in the station or precinct house)
- Line-of-duty safety practices (such as safe driving, hazard awareness, and personnel accountability)
- Enforcement of PPT use
- Physical fitness and wellness promotion and testing.

Many participants candidly acknowledged that they were struggling with deficiencies in these areas and were seeking reliable solutions. Participants spoke about programs and measures, often ad hoc, that they had pursued to remedy those deficiencies, often with limited success. At the same time, many participants pointed to models of good practice in other agencies and services.

Participants typically learn about examples of good practices at professional meetings and or by reading professional publications.

However, information on best practices is largely based on anecdotal evidence and does not provide a reliable guide for safety management. For example, there was little uniformity in participants' descriptions of the key attributes of a successful or unsuccessful PPE enforcement or physical fitness program, other than mentioning broad categories such as "leadership" and "funding." This response is not surprising: To assess and compare the performance of personal protective equipment in a laboratory setting is easier than attempting to assess and compare the performance of organizations in a community setting. The need for further study and benchmarking of organizations' behavior and practices is not trivial: As we have seen, personal protection practices are as critical to responder health and safety as personal protection equipment.

Government enforcement of federal and state occupational health and safety regulations in industry has resulted in efforts to better understand variations in safety program performance among companies. As firms have sought to manage financial and legal risks and boost employee morale and retention, various benchmarking efforts have been conducted to identify critical organizational variables—such as leadership, communications, teamwork, and morale—and more rigorously measure and compare those variables among peer firms. The absence of federal OSHA regulation of state and municipal agencies has created a disincentive for such efforts in the emergency response community. However, developing the ability to rigorously document and compare practices among emergency response organizations could tease out critical variables that are essential for improving responder health and safety and establish reliable benchmarks against which organizations can be compared.

POLICY ISSUES FOR THE FUTURE

In addition to uncovering differences in how organizations and individuals view occupational hazards and personal protection needs, the discussions with participants also elicited fundamental differences of opinion on important health and safety policy issues for the emergency response community. In this section, we touch on some of those policy debates.

Many policy issues are complex and pose challenging questions. However, one broad question permeates most of these issues: What is the proper balance between distributed and centralized decisionmaking? In the United States, emergency response is handled in a highly decentralized, grassroots manner. Solutions to problems, such as problems concerning personal protection, largely have been left to the local departments, or even the individual responder, and are heavily driven by the free market. In the post–September 11 environment, the perceived threats to responders' health and safety have become more varied and complex and are on a scale previously unimagined. As a result, some community members raised fundamental questions about the adequacy of current decisionmaking strategies and put forth ideas (summarized in Table 9.2) that suggested the need for a more top-down, directed, or hierarchical approach to addressing fundamental policy issues.

Personal Protective Technology Research and Development

RAND's discussions with the emergency responder community revealed that there are a number of impediments to, and resulting gaps in, PPT research and development that limit progress in reducing injuries to, and improving the capabilities of, the emergency responder workforce.

Table 9.2

Key Policy Areas and Issues Raised by the Emergency Responder Community

Policy Areas	Specific Issues
PPT research and development	• Research should be more strategic and multidimensional, including more fundamental, long-term research • Greater emphasis on ensembles is needed • R&D should address response activity rather than services • Decentralized market limiting innovation and purchasing power should be addressed
Discretion in personal protection decisionmaking	• Expanding role of emergency responders and improved hazard assessment warrant increased attention to activity-specific tailoring of protection
PPT standards for emergency medical services and law enforcement	• EMS and police communities need dedicated personal protection, safety, and standardization efforts
PPT performance assessment	• Reliable and objective equipment performance assessments need to be developed
PPT standardization and interoperability	• Mutual-aid agreements and extended operations should be facilitated by enhanced standardization and interoperability
The role of risk in emergency response	• Examine emergency responders' perceptions of and their responses to risks inherent in emergency response • Promote efforts to decrease risk through improved information management, clarified protocols, and improved equipment

A common complaint we heard was that the emergency responder community, while large in numbers of professionals and volunteers, has limited purchasing power compared with industrial users of similar technologies. This limited buying power is the result of the decentralized market for PPT that the emergency responder community constitutes and tight municipal budgets. The PPT market is so decentralized because municipalities, acting independently, are the principal purchasers of PPT. Many participants noted that municipal personal protection equipment and training budgets were very tight because of widespread weakness in the U.S. economy and unprecedented budget shortfalls experienced by many city and state governments. In addition, resources that were promised by the federal government had yet to be appropriated.

Limited purchasing power may also limit the influence of the emergency responder community on research and development directions. Many participants felt that, as a result, R&D is driven largely by the priorities of industry. Widespread dissatisfaction with radios issued for firefighting was cited as one problem resulting from PPT not being developed specifically for emergency responders' needs. Similarly, the historic lack of a substantial mainline law enforcement market for PPT was blamed for the absence of equipment designed specifically for law enforcement.

To add to the decentralized PPT market, much of the PPT research and development has been conducted by equipment and services suppliers. These entities are also decentralized and often focus on a narrow segment of the development chain, further fragmenting the PPT development and diffusion process. For example, respirator manufacturers develop respiratory protection, garment makers produce protective clothing, and various other businesses produce helmets, boots, radios, and sensors, generally with little coordination among these groups. Even for a single component, development may be spread among several entities. Protective garments, for example, often evolve from development at three levels, with innovations at each level occurring in a separate field: chemical companies develop new materials, textile mills combine these materials into fabrics, and garment manufacturers focus on equipment design and performance. One important result of this fragmented supplier base is that development of protective technologies and the standardization and certification process have been focused primarily on discrete components rather than on the entire ensemble.

Another challenge in PPT R&D is that emergency responders have very diverse protection needs. The hazards they face can vary according to a number of factors, including the branch of service, the size of the community, or even the time of day. Emergency personnel must work in a wide variety of incident environments that have a unique mix of hazards and that are often dynamic and unpredictable. The hazards associated with illegal drug laboratory responses, for example, are of particular concern to law enforcement agencies. As a result, multiple-use components (i.e., those offering protection against a wide spectrum of risks) are highly valued by emergency responders. A desire for integrated, multiple-risk protection designs is further driven by cost and logistic factors. Smaller organizations with limited equipment and training budgets, especially volunteer organizations, cannot handle complex equipment logistics (e.g., storage, transportation and maintenance). Again, the fragmented nature of personal protection development and demand tends to reduce market incentives to develop universal solutions.

RAND's discussions with the emergency response community brought to light several broader impediments to PPT research and development:

- **Cost**. Research and development efforts aimed at total ensemble solutions are complex. For example, participants mentioned the difficulty of conducting applied research, creating standards, and establishing certification procedures for an entire firefighter ensemble and its various wearing positions, such as standing, crouching, and lifting. Financial resources may need to be directed at research that takes a total-ensemble approach.

- **Short time horizons**. In responding to demand driven by a fragmented municipal-level market, most R&D efforts are focused on short-term incremental advances. Research aimed at innovations desired by the entire emergency response community that would have far-reaching effects (e.g., ultra-lightweight and breathable protective materials) is very limited and not sustained.

- **Service orientation**. Most PPT R&D is focused on serving the needs of a specific service. However, community representatives suggested that their personal protection concerns often cross service boundaries. For example, members of every service expressed the desire for better chemical and biological detection capabilities and better biological- and chemical-protective capabilities of garments and respirators. A service-specific orientation to PPT R&D thus contributes to the fragmentation of R&D and may also leave critical R&D gaps: Little R&D is directed toward the needs of emergency medical personnel, and far less is being directed toward the needs of other personnel at major incident sites, such as sanitation workers, public works personnel, and construction trades workers.

Addressing these shortcomings may require a new approach to PPT research and development, namely the initiation of more strategic, top-down priority setting. The personal protection research agenda that would result from this approach would have diverse goals. The research would be balanced between short-term efforts addressing the performance shortcomings of discrete components, such as boots and helmets, and longer-term basic research aimed at less-specific, but not less-important, goals. One example is addressing ways to improve protection against toxic chemicals and communicable diseases. Similarly, such research needs to address component integration: Participants repeatedly stated that it is critically important that equipment components be compatible—i.e., interface properly—and that they do not detract from the overall performance of other equipment or the ensemble. Finally, personal protection research would be driven by hazard-, incident-, and service-oriented perspectives.

Discretion in Personal Protection Decisionmaking

One of the major concerns to emerge from the discussions is the increasing variety of response scenarios and specialized tasks that emergency responders must undertake. This trend has raised important policy questions about the extent to which emergency responders should have more-specialized or risk-specific PPT alternatives rather than all-purpose personal protection options.

A related issue concerns determining the appropriate decisionmaking level for assessing the risks that emergency responders face and choosing the personal protection options to address those risks. For structural firefighting, for example, much of the risk assessment is essentially conducted at the national level. Decisions governing the selection of protective equipment and protocols for the use of that equipment are made by national-level institutions and are implemented through communitywide standards (e.g., NFPA 1971). These standards have favored a universal approach to personal protection—for example, use of an SCBA for all respiratory-protection needs. These standards have become so widely accepted that they have been implemented by most fire departments in the country and are referenced in the legislation of many states.

As a consequence of having a nationally imposed standard that calls for a single protective ensemble for all responses, firefighters across the nation use similar protective equipment in similar ways. The result is a decisionmaking model in which risk assessment, at least in terms of the thermal-protective ensemble, has been pushed "upstream" of departments or individuals. Such a model has important trade-offs, many participants observed.

An advantage to using this model is that standards reduce the need for individuals and departments to assess risks at an incident scene, and they provide clear guidance for firefighter protection. The process of assessing risk and evaluating protection needs is a difficult and complex one, and the majority of fire departments do not have the qualifications and resources to carry it out.

A disadvantage of this model is that it reduces risks and protection needs to a common conservative (i.e., maximum protection) baseline. As one participant noted, "A fire in Boston is about the same as a fire in Seattle," but variations in climate, building construction, city layout, and industrial-residential mix, as well as the capabilities and resources of individual departments, can mean different risks and protection needs from one location to another. Similarly, standards exist for only a single type of protective ensemble, and those standards are designed to protect against the most serious hazard—fire. Not given consideration are other types of incidents to which firefighters often respond, such as vehicle collisions, medical emergencies, structural collapses, and brush fires, which result in exposures and risks that are frequently different from those in a structural fire.

In some cases, risk-specific protection can be partially achieved through customer-specified design of the protective gear. However, in many incidents, responders may disregard standards and informally assemble "risk-specific" ensembles through "mixing and matching" or layering of components. Some departments and firefighters already are using alternative gear, such as ranger boots, battle dress uniforms (BDUs), and lightweight gloves, on an informal basis. Informal adaptations of protective gear typically involve individual decisionmaking and may be inconsistent with the gear's intended design. As a result, this practice may leave firefighters underprotected.

Instituting a single standard has been very effective for raising firefighting organizations nationwide to a common baseline that offers a very high level of protection. However, hazard- and risk-assessment capabilities are improving with better training and diagnostics, potentially enabling departments and individuals to select more-appropriate levels of protection based on the known risks rather than a level of protection prescribed by existing national standards. This suggests that the national risk-assessment and standard-setting model used in the fire service could be improved by allowing for more location-specific and incident-specific information to be used in determining protection needs and by giving firefighters a range of protective options.

Formal movement toward sanctioning the use of incident-specific gear has occurred with the introduction of NFPA garment standards for EMS use (National Fire Protection Association, 1997) and for USAR use (National Fire Protection Association, 2001c). These standards specify design, performance, and testing criteria for protective clothing to be used for particular activities. While the introduction of such standards is an important first step, an issue that has yet to be addressed is the need for criteria for deciding when different types of protective ensembles should be used.

Personal Protective Technology Standards for the Emergency Medical and Law Enforcement Services

A key component of any personal protection strategy is finding ways to maximize PPT use and compliance. Findings from the discussions indicate that PPT use and compliance have been high-priority concerns, and that compliance in certain areas has improved considerably in recent years, particularly in firefighting. Use of SCBAs during fire attacks, for example, is nearly ubiquitous among the departments with whom we met. Similarly, nearly all firefighters wear NFPA-compliant turnouts. Increased PPE compliance has been a hard-won struggle, and several departments echoed the sentiment of one fire chief, who noted that, "If you had asked five years ago, the answer would have been different."

Increased PPE compliance in the fire service was attributed to a number of factors, including promulgation and updating of equipment and practice standards; availability of more comfortable and ergonomically correct equipment; improved compliance training; increased awareness of studies demonstrating the effectiveness of protection use in reducing injuries; more stringent certification requirements that have driven noncompliant gear out of the market; and more strict enforcement at the department, state, and federal levels.

The extent of PPT use varies significantly across the emergency response community. Specifically, PPT use in emergency medical service response and law enforcement was repeatedly cited as being far behind PPT use in firefighting. One reason for the lower levels of PPT use in emergency medical services and law enforcement was the lack of clearly defined and well-accepted standards for PPT design, performance, and use in emergency response. For example, the majority of emergency medical service responders noted that they prefer disposable protection (e.g., sleeves, gowns, masks, gloves), for which few standards exist outside of the standards for clinical (e.g., hospital) and industrial settings. As one participant put it, "We want EMS to be like firefighting, where all the junk is off the market." Some law enforcement and EMS organizations have sought to circumvent this shortcoming by adopting protective equipment and practices certified by outside agencies, such as NFPA, NATO, ISO, and OSHA, whenever possible.

Efforts to enhance responder safety in law enforcement and the emergency medical services have been undertaken by federal agencies, professional associations, and labor unions. However, neither law enforcement nor EMS receives the level of guidance and support that the fire service receives from the federal government (through the U.S. Fire Administration) and a dedicated professional body (NFPA). Law enforcement benefits from the support of NIJ, but it has no NFPA analog. EMS organizations, on the other hand, benefit from the guidance of the National Association of Emergency Medical Technicians, but they have no support from a dedicated government agency. The shortcomings in the level of formal guidance and support that the emergency medical and law enforcement services receive present formidable hurdles to improving responder safety in areas such as equipment design, testing, certification, and procurement; occupational health and safety research; compliance and enforcement; and safety education, training, and communications.

Personal Protective Technology Performance Assessment

Numerous participants mentioned identifying, evaluating, and selecting protective technologies as areas of the acquisition process that needed improvement. While NFPA, NIJ, and other design and performance standards ensure a basic level of functionality and protection, distinguishing among the wide variety of certified gear within each equipment class is not straightforward. Most responder organizations resort to informal, ad hoc personal protection technology evaluation and information-gathering and analysis efforts because they lack access to reliable information sources on PPT performance to inform their procurement decisions.

Given this situation, the creation of an objective, third-party assessment capability would greatly facilitate PPT evaluation and acquisition decisions. A few departments have formal in-house evaluation capabilities, and some departments hire outside consultants to perform this function, but these options are available only to larger, more-wealthy organizations. As discussed in Chapter Eight, the National Institute of Justice took an important step toward performance evaluation by producing a resource guide to assist with the selection of chemical and biological agent protection for emergency responders (National Institute of Justice, 2002), one of a series of NIJ guides on technologies for emergency responders. The NIJ guides provide performance evaluations for commercially available equipment based on a suite of selection criteria and vendor-supplied performance data. The guides represent an important contribution, given that many RAND participants indicated that they lacked detailed knowledge about PPT, particularly for chemical and biological response.

The emergency response community would benefit from similar third-party information and performance assessment of PPT for firefighting, community policing, medical response, and other conventional responder activities. Reliable third-party PPT performance assessment would in particular facilitate the PPT decisionmaking process regarding new technologies for the nation's smaller emergency response organizations.

Personal Protective Technology Standardization and Interoperability

Although emergency response often requires mutual aid among responder organizations, acquisition of PPT is rarely coordinated between services and jurisdictions to ensure interoperability (interchangeability) and sharing of equipment (for instance, sharing of respirator components). PPT interoperability has been a subject of discussion in the emergency responder community for many years. Problems with a lack of interoperability of PPT and a lack of uniform PPT training, maintenance, and use protocols for responders at the scene of the World Trade Center attacks and other terrorist incidents have raised the importance of this subject as a policy matter (Jackson et al., 2002).

Interoperability and standardization may be addressed from the bottom up through greater interagency coordination of acquisitions and training. Among the local agencies that participated in the RAND discussions, this coordination was not a high priority. Moreover, the costs of transitioning to a new technology combined with agencies' historical allegiances to specific equipment and suppliers are substantial impediments to change.

An alternative is to pursue interoperability from the top down through, among other strategies, the promulgation of federal uniform design standards or purchasing arrangements. In the communications arena, the federal government has for many years encouraged and supported the development and diffusion of 800-MHz trunked radio systems to facilitate interagency communications, response coordination, and mutual assistance. In 1998, the Department of Defense and the Department of Justice founded the InterAgency Board for Equipment Standardization and InterOperability which, as a first step, has developed a national Standardized Equipment List for use by responder agencies and organizations in preparing for and responding to weapons of mass destruction terrorism (InterAgency Board for Equipment Standardization and InterOperability, 2001).[1]

The nation's goal of improving homeland defense capabilities further suggests that efforts to promote PPT standardization and interoperability should be a community priority. In addition to facilitating mutual aid and equipment sharing at large-scale events, standardization may help to

- facilitate potential technology transfer and equipment sharing between civilian and military organizations

- promote economies of scale and lower costs for equipment acquisitions and logistics

---

[1]The Standardized Equipment List recommends equipment types; in most cases, it does not prescribe specific proprietary brands or designs.

- simplify PPT evaluation and acquisitions decisionmaking

- focus PPT education and training efforts.

However, the low priority for PPT interoperability at the local level and problems with its implementation (as illustrated by the lagging adoption of 800-MHz trunked communications systems) suggest that substantial financial and other incentives will have to be provided to local authorities to facilitate the transition to standardized PPT and to realize the benefits noted earlier.

The Role Risk Plays in Emergency Response

The inherent risks in emergency response are heightened by responders' personal and professional commitments to render assistance, characterized by the credo "Risk a life to save a life." Many participants noted that the high-stakes nature of the profession was part of what makes it attractive to potential recruits. The specter of weapons of mass destruction has raised the stakes even higher. Several participants also observed that emergency response doctrine has become more proactive in recent years, given the greater need, for example, to aggressively confront terrorists. More-aggressive postures have been supported by better protective technologies. At the same time, improvements in PPT, many senior personnel argued, have encouraged greater risk-taking by emergency responders.

Difficulties in formulating policy regarding risk-taking in emergency response partly stem from the lack of specific information about the effect of responder behavior on health and safety. While a significant volume and variety of injury data have been collected, an area in which occupational health and safety surveillance is particularly incomplete is in the role of emergency responder behavior. Participants raised questions in a number of areas about the impact of responder behavior on safety, including the following:

- What are the merits of wearing turnout gear during nonfire responses?

- How great is the need for responder rehabilitation at extended responses?

- What is the impact of responder fitness on managing physical stress?

- What are the costs versus the benefits of emergency vehicles traveling at high speed in "lights and sirens" scenarios?

- What constitutes improper or overextended PPT use?

More-extensive data collection and dissemination, especially as they relate to police and emergency medical service responders, would help guide personal protection R&D, education, and training activities. It could also be used to modify responder behavior, said many participants, by heightening responders' awareness of the risks associated with particular decisions.

At the command and department level, questions about risk arose in the context of the extent to which emergency response should be "offensive" (e.g., acting to stabilize a situation as quickly as possible) or "defensive" (e.g., retreating from a hazardous scene). One fire service representative described how his department approached protection through a combination of procedures, engineering, and, in some cases, simply staying away, noting that "every hazard will self-mitigate eventually." Many participants noted that fire services elsewhere in the world tend to assume a more defensive stance from the start of an event, and law enforcement representatives spoke of strict policies that their agencies were implementing governing high-speed pursuits.

Since the loss of more than 400 responders in the collapse of the World Trade Center towers, the issue of risk has been uppermost in the minds of both commanders and rank-and-file personnel. The anthrax attacks also introduced heightened cautiousness across the emergency response community. Police commanders, for example, openly questioned the merit of sending their personnel into zones with unknown hazards given that most personnel lack critical personal protection equipment and training.

The questions raised in this report about the current and potential hazards that emergency responders face and the changing nature of the emergency responder profession point to fundamental policy issues that must be addressed. These issues merit focused discussion across the entire emergency responder community as the United States enters a new era in which emergency responders must be fully prepared to meet not only the challenges that routine emergencies present but also new challenges emerging from an increasingly unpredictable environment.

DISCUSSION PARTICIPANTS

FIRE SERVICES

Arlington County (VA) Fire Department

D. G. Bingham, Captain/Technical Rescue

John Delaney, Firefighter/Hazmat Technician

Kenneth Johnson, Captain/Fire and EMS

George Lyon, Battalion Chief/Technical Rescue Coordinator

Boeing Fire Department (Seattle, WA)

David Cook, Chief

Steven Foley, Hazmat Response/Security and Fire Protection

Gary Gordon, Toxicologist/Security and Fire Protection

Boston Fire Department

Paul Christian, Chief

Chicago Fire Department

John Eversole, Coordinator of Hazardous Materials

City of Austin (TX) Fire Department

Tyler Anderson, Assistant Director

David Beardon, Battalion Chief/Safety Officer

Charles Catt, Division Chief

Jim Evans, Assistant Fire Chief

City of Austin (TX) Fire Department—cont.

Phil Jack, Division Chief

Paul Maldonado, Fire Marshall

Dayton (OH) Fire Department

Rennes Bowers, Captain/WMD Emergency Operations Planning

Lacey Calloway III, Assistant Chief

Larry Collins, Director and Chief

Joe Hoying, Captain/Safety Officer

Elk Grove Community Services District (CA) Fire Department

Keith Grueneberg, Deputy Chief

Richard Holmes, Battalion Chief/Special Operations

John Michelini, Battalion Chief

Fairfax County (VA) Fire & Rescue Department

Glenn Benarick, Deputy Chief/Fire Prevention

Dean Cox, Captain/Resource Management

Dewey Perks, Battalion Chief/Special Operations and USAR Task Force Leader

Fire Department of the City of New York

Joseph Governale, Captain/Decontamination, Inspections and Safety

John Norman, Chief/Special Operations

Hillsboro (OH) Fire Department

Jerry Powell, Chief

Los Angeles City Fire Department

Frank Borden, Assistant Chief (Retired)/Urban Heavy Rescue Task Force

Dean Cathy, Assistant Chief/Director of Bureau of Emergency Services

Lubrizol Corporation (Deer Park, TX)

Stephen Greco, Safety Supervisor

Lyondell-Citgo Refining, LP (Houston, TX)

Pete Greco, Assistant Fire Chief/Emergency Management Coordinator

Oakland (CA) Fire Department

Mark Hoffman, Captain/Safety Officer

Sid King, Captain/Hazardous Materials

Bill Wittmer, Assistant Chief/Special Operations

Oklahoma City Fire Department

James Blocker, Major/EMS Quality Assurance Officer

Bryan Heirston, Battalion Chief/Safety Officer

Bill Williams, Major/Hazardous Materials

Okmulgee (OK) Fire Department

Bob Hartridge, Chief

Olympia (WA) Fire Department

Pat Dale, Assistant Chief

Greg Wright, Assistant Chief/Risk Management and EMS

Pasadena (CA) Fire Department

Ernest Mitchell, Chief

Pawtucket (RI) Fire Department

James Condon, Chief

Phoenix Fire Department

Dawn Bolstad-Johnson, Industrial Hygienist

Ron Cobos, Captain/Special Operations

Mark Delima, Captain/Hazardous Materials Program Manager

Timothy Durby, Research and Development Program Manager/Resource Management Division

Kevin Roche, Resource Management Administrator

Pittsburgh Bureau of Fire

Thomas Airesman, Captain/Communications

David Borgese, Station Captain

Fred Childs, Firefighter/Union Occupational Health and Safety Committee Chairman

Arthur George, Assistant Chief/Operations

David Grady, Firefighter/Training Academy Instructor

Peter Micheli, Jr., Chief

Robert Modrak, Deputy Chief/Administration

Robert Walker, Lieutenant/Fire Prevention

Rural/Metro Fire Department (Scottsdale, AZ)

Daniel Bunce, Battalion Chief

Brian Dickes, SCBA Technician

Joseph Early, Captain/Training

Kore Redden, Compliance Officer

Salt Lake City Fire Department

Raleigh Bunche, Battalion Chief/Safety and Wellness

San Antonio Fire Department

David Coatney, Captain/Safety Officer

San Francisco Fire Department

Karl Hillyard, Paramedic Captain/Special Operations

Paul Jones, Assistant Chief/Safety Division

James McCaffrey, Paramedic Captain/Special Operations

Robert Navarro, Section Chief/Special Operations

Santa Monica (CA) Fire Department

Jim Hone, Assistant Chief/Fire Marshall

Seattle Fire Department

James Fosse, Assistant Chief/Medical and Safety

Rick Newbrey, Lieutenant/Medical Services

Edwin Peterson, Lieutenant/Hazardous Materials

Nick Ponce, Lieutenant/Commissary

Geoff Wall, Captain/Support Services

A. D. Vickery, Deputy Chief/Special Operations

Sierra Madre (CA) Volunteer Firefighters Association

Bill Messersmith, President

Texas City (TX) Fire Department

Gerald Grimm, Chief

Oseeg Sowell, Captain/Executive Officer

Tracy (CA) Fire Department

Pete Luckhardt, Engineer

Tulsa (OK) Fire Department

Randy Brasfield, Fire Training Officer

Michael Mallory, Fire Protection Engineer, Safety and Engineering Branch

Wilmington (NC) Fire Department

R. V. Jordan, Battalion Chief/Special Teams Coordinator

Ron Little, Captain

Worcester (MA) Fire Department

Walter Giard, Division Chief

John Griffin, Lieutenant

Robin Mitchell, Firefighter

EMERGENCY MEDICAL SERVICES

City of Austin/Travis County (TX) Emergency Medical Service

Gordon Bergh, Assistant Director/Operations

Christian Callsen, Jr., Senior Division Commander

Richard Herrington, Executive Director

Emergency Medical Services Authority (Tulsa, OK)

Aaron Howell, Director/Operations

Pittsburgh Bureau of Emergency Medical Services

Robert Farrow, Division Chief

LAW ENFORCEMENT

Baytown (TX) Police Department

D. W. Alford, Lieutenant

Boston Police Department

Bill Good, Chief/Administrative Services

Paul Joyce, Superintendent/Special Operations

Dayton (OH) Police Department

John Bardun, Lieutenant/Special Operations

Randy Beane, Lieutenant/SWAT

Dennis Chaney, Sergeant/Bomb Squad

Roy Ewing, Detective/Bomb Squad

Carol Johnson, Detective

Bob Murchland, Detective

Hillsboro (OH) Police Department

John Salyer, Officer

Houston Police Department

Steve Connor, Sergeant/Bomb Squad

Richard Kleczynski, Lieutenant/Tactical Operations Division

Michael Walker, Captain/Tactical Operations Division

Los Angeles County Sheriff's Department

Heidi Clark, Sergeant/Arson, Explosives Detail

Charles Heal, Captain/Special Enforcement Bureau

Metropolitan Police Department (Washington, DC)

Jeffrey Herold, Lieutenant, Special Operations Division

Muskogee (Creek) Nation Tribal Police (Okmulgee, OK)

Prentiss Berryhill, Assistant Chief

Washington Cummings, Chief

New York City Police Department

Kevin Devine, Hazmat Specialist, Emergency Service Unit

Gerard DiMuro, Sergeant, Quartermaster Section

Thomas Dowd, Sergeant/Fiscal Coordinator, Emergency Service Unit

Richard Florentino, Lieutenant, Quartermaster Section

Patrick Pogan, Detective, FBI-NYPD Joint Terrorism Task Force

Nicholas Russo, Sergeant, Disorder Control Unit

Dani-Margot Zavasky, Counterterrorism Bureau

Oklahoma City Police Department

Steve McCool, Captain/Departmental Safety Officer

Pawtucket (RI) Police Department

Paul King, Captain

Stephen Ormerod, Commander

Phoenix Police Department

Mike DeBenedetto, Lieutenant/Operations Support

Pittsburgh Bureau of Police

Charles Moffatt, Deputy Chief

Linda Rosato-Barone, Commander/Chief of Staff

Providence (RI) Police Department

Napolean Brito, Detective Sergeant

Martin Hames, Major

Vincent Mansolillo, Detective Sergeant

Richard Sullivan, Colonel/Chief of Department

Rhode Island State Police

John Leyden, Jr., Major

Glenn Skalubinski, Lieutenant

Gary Tremil, Captain

Richmond (CA) Police Department

David O'Donnell, Lieutenant, SWAT

Joseph Samuels, Chief

Doug Seiberling, Captain, SWAT

Salt Lake City Police Department

Scott Folsom, Assistant Chief, Investigative Bureau

Carroll Mays, Commander, Liberty Patrol

Santa Monica (CA) Police Department

Gary Gallinot, Commander/Office of Administrative Services

Seattle Police Department

Dan Bryant, Assistant Chief, Criminal Investigations Bureau

James Fitzgerald, Lieutenant/Training, SWAT

Ted Jacoby, Captain/Communications

Andy Tooke, Lieutenant/Commander, Special Assignments Unit

Texas City (TX) Police Department

Ronald Berg, Captain, Administration

Driscoll R. Young, Sergeant

Tulsa (OK) Police Department

Dennis Larson, Captain/Commander, Bomb Squad

Lawrence McCoy, Captain, Special Operations Division

Wilmington (NC) Police Department

Randy Pait, Captain/Patrol Commander, Community Policing Division

EMERGENCY MANAGEMENT

Los Angeles County Emergency Operations Bureau

Michael Grossman, Captain/Director, Emergency Operations Bureau

Jeffrey Marcus, Battalion Chief, Los Angeles City Fire Department

Ronald Watson, Battalion Chief, Los Angeles County Fire Department

New Hanover County (NC) Department of Emergency Management

Michael George, Emergency Management Specialist

Oklahoma City Office of Emergency Management

John Clark, Director

Ronnie Warren, Deputy Director

TECHNOLOGY AND SERVICES PROVIDERS

E. I. DuPont deNemours and Company (Richmond, VA)

Richard W. Blocker, Jr., Fire Service Segment Leader, Nomex Personal
Protection Solutions

Dave Martin, Life Protection Division

Dale Outhous, Global Business Manager, Protective Apparel

Jim Ransom, Jr., Kevlar Business Development Manager

Rich Young, Senior Research Chemist

Jim Zeigler, Research Associate, Nonwovens

Emergency Planning and Response Consulting (Wilmington, NC)

Jeff Babb, Principal

Industrial Scientific Corporation (Oakdale, PA)

Kent McElhattan, President and Chief Executive Officer

Richard Warburton, Manager, Research and Development

International Personal Protection, Inc. (Austin, TX)

Jeffrey Stull, President

Lion Apparel (Dayton, OH)

Don Aldridge, Vice President, Research and Development

Nick Curtis, Vice President, Product Development

John Neal, Major Accounts Manager

Frank Taylor, Director, Textile Merchandising

Mine Safety Appliances (Pittsburgh, PA)

Kenneth Bobetich, Product Group Manager/Air-Purifying Respirators

Ron Herring, Director of Marketing

John Kuhn, Product Engineering Manager/Supplied-Air Respirators

Richard Moore, Product Engineering Manager/Air-Purifying Respirators

Michael Rupert, Product Group Manager/Head, Eye, Face, and Hearing Protection

National Academy of Emergency Medical Dispatch (Salt Lake City, UT)

Jeffrey Clawson, Chair, Board of Certification

TotalFire Group (Dayton, OH)

Bill Grilliot, President and Chief Executive Officer

Mary Grilliot, Vice President and Chief Operating Officer

Underwriters Laboratories, Inc. (Research Triangle Park, NC)

Steven Corrado, Engineering Group Leader

Daniel Ryan, Associate Managing Editor

Gregg Skelly, Senior Engineering Associate

RESEARCH, POLICY, AND PROGRAMS

Building and Fire Research Laboratory, National Institute of Standards and Technology (Gaithersburg, MD)

David Evans, Fire Research Division

Center for Emergency Response Technology, Instruction and Policy, Georgia Institute of Technology

Thomas Bevan, Director

Center for Research on Textile Protection and Comfort, North Carolina State University

Roger Barker, Director and Burlington Chair in Textile Technology

Department of Public Management, John Jay College of Criminal Justice

Charles Jennings, Professor

International Association of Firefighters (Arlington, VA)

Richard Duffy, Director

Andy Levinson, Health and Safety Specialist

Maryland Fire and Rescue Institute, University of Maryland

Steven T. Edwards, Director

National Fire Protection Association (Quincy, MA)

Bruce W. Teele, Senior Fire Service Safety Specialist, Public Fire Protection Division

Gary O. Tokle, Assistant Vice President, Public Fire Protection Division

National Terrorism Preparedness Institute, St. Petersburg College

David Puckett, Deputy Director, Technology

Office of Law Enforcement Standards, National Institute of Standards and Technology (Gaithersburg, MD)

Alim A. Fatah, Program Manager, Chemical Systems and Materials

Kathleen M. Higgins, Director

Philip J. Mattson, Chem/Bio Program Integration, Defense Services

Oklahoma City Memorial Institute for the Prevention of Terrorism

Jim Gass, Plans and Special Projects Officer

Brian Houghton, Director of Research

Dennis Reimer, Director

Division of Textiles and Clothing, University of California, Davis

Gang Sun, Associate Professor

U.S. Army Soldier and Biological Chemical Command (Natick, MA)

John Gassner, Director, Supporting Science and Technology Directorate

William Haskell, Technical Program Development

Robert Kinney, Director, Individual Protection Directorate

U.S. Navy Clothing and Textile Research Facility (Natick, MA)

Joe Giblo, Biomedical Engineer

Sue Reeps, Director, Protective Clothing Division

Harry Winer, Textile Toxicologist

DISCUSSION PROTOCOL

PERSONAL PROTECTIVE EQUIPMENT

- In your view, what are the activities and situations in which the responders in your organization are at the greatest risk of injury or illness? Why are these the most risky situations?

- Where are the weakest links in your protective equipment?

- What are your priorities for acquiring personal protective equipment now or in the future?

- Over the past few years, have emergency responders in your organization suffered any casualties that resulted from shortcomings in the design or use of PPE? How has this affected your thinking about PPE needs, training, or information?

- In your view, what are the currently available or emerging innovations that would be most beneficial for increasing the protection of your organization's members and/or enhancing the capabilities of individual emergency responders?

TECHNOLOGIES RELATED TO PERSONAL PROTECTION

- What types of hazard monitoring and assessment technologies do you use in your operations? How does this information help you select and use personal protective equipment? What innovations in this area would make the greatest difference to your organization?

- Do you perceive that there is a need for more tailored PPE options to provide more-task-specific PPE based on characteristics of

    - particular types of responses?

    - particular activities at a given response?

- What communications innovations would make the greatest difference to your organization with regard to protecting responders?

PERSONAL PROTECTIVE TECHNOLOGY MANAGEMENT AND TRAINING

- How does your organization manage the maintenance, repair, and retirement of personal protective equipment? What information or innovations would make maintaining and assuring the performance of your PPE easier or more effective?

- Proper training and information are critical for maximizing the effectiveness of PPE. Is the current situation adequate? What options do you have? What innovations would improve information and training regarding PPE use?

INTERAGENCY COORDINATION

- With respect to personal protective equipment, do you coordinate with other districts or emergency response agencies regarding PPE interoperability, supply, training, or enforcement?

PERSONAL PROTECTION MARKET

- How do you assess your department's PPE needs?

- How do you evaluate personal protection technologies and practices? How do you acquire them? Are there ways to improve the flow of information?

- What factors influence your decisions as to which personal protective technologies to acquire (e.g., cost, regulatory requirements, usefulness, quality, compatibility with existing equipment)?

- How does certification affect PPE acquisition and use? What are your experiences and feelings regarding certification of PPE?

THE FUTURE ENVIRONMENT

- How do you foresee changes in the mission/role of your organization over the next several years impacting your personal protection needs? What are the risk and personal protection implications?

Bureau of Labor Statistics. (2002). *Survey of Occupational Injury and Illness: States of California, Maine, New Jersey, New York, North Carolina,* Washington, D.C.: Bureau of Labor Statistics (available at http://www.bls. gov/iif/oshstate.htm; last accessed October 2002).

Bureau of Labor Statistics. (2003a). *Occupational Employment Statistics Survey by Occupation, 2001,* Washington, D.C.: Bureau of Labor Statistics (available at http://www.bls.gov/oes/2001/oes_29He.htm; last accessed March 2003).

Bureau of Labor Statistics. (2003b). *Survey of Occupational Injuries and Illnesses,* database report, obtained via personal communication from Boston–New York regional office staff, March 2003.

Clarke, C., and Zak, M. J. (1999). "Fatalities to Law Enforcement Officers and Firefighters, 1992–1997," *Compensation and Working Conditions,* Bureau of Labor Statistics, pp. 3–7.

Dower, John M., Richard W. Metzler, Frank M. Palya, Jeff A. Peterson, and Molly Pickett-Harner. (2000). *NIOSH-DOD-OSHA Sponsored Chemical and Bio-logical Respiratory Protection Workshop Report,* Pittsburgh, Pa.: National Institute for Occupational Safety and Health.

Dwyer, Jim, Kevin Flynn, and Ford Fessenden. (2002). "9/11 Exposed Deadly Flaws in Rescue Plan," *New York Times,* July 7, 2002, p. A1.

Heightman, A. J. (2000). "EMS Workforce," *Journal of Emergency Medical Services,* 25, pp. 108–112.

Hickman, Matthew J., and Brian A. Reaves. (2001). *Local Police Departments, 1999,* Washington, D.C.: Bureau of Justice Statistics.

Hickman, Matthew J., and Brian A. Reaves. (2003). *Local Police Departments 2000,* Washington, D.C.: Bureau of Justice Statistics.

InterAgency Board for Equipment Standardization and InterOperability. (2001). *2000 Annual Report,* Arlington, Va.: InterAgency Board for Equipment Standardization and InterOperability.

Jackson, Brian A., et al. (2002). *Protecting Emergency Responders: Lessons Learned from Terrorist Attacks*, Santa Monica, Calif.: RAND, CF-176-OSTP (available at http://www.rand.org/publications/CF/CF176/).

Karter, Michael J., Jr. (2000). *Patterns of Firefighter Fireground Injuries*, Quincy, Mass.: National Fire Protection Association.

Karter, Michael J., Jr. (2001). *U.S. Fire Department Profile Through 2000*, Quincy, Mass.: National Fire Protection Association.

Langan, Patrick A. (2001). *Contacts Between Police and the Public*, Washington, D.C.: Bureau of Justice Statistics.

Maguire, B. J., K. L. Hunting, G. S. Smith, and N. R. Levick. (2002). "Occupational Fatalities in Emergency Medical Services: A Hidden Crisis," *Annals of Emergency Medicine*, 40, pp. 625–632.

McKinsey & Company. (2002). *Increasing FDNY's Preparedness*, New York, N.Y. (available at http://www.nyc.gov/html/fdny/html/mck_report/toc.html; last accessed December 2002).

National Association of Emergency Medical Technicians. (2002). Telephone survey of state and territorial EMS Offices, Clinton, Miss., March 1, 2002 (obtained via personal communication from Paul M. Maniscalco, February 2003).

National EMS Memorial Service. (2002). *Notices of Line of Duty Death*, Oilville, Va.: National EMS Memorial Service (available at http://nemsms.org/notices.htm; last accessed December 2002).

National Fire Protection Association. (1995–2000). *Firefighter Injury Reports*, Quincy, Mass.: National Fire Protection Association.

National Fire Protection Association. (1995–2001). *Firefighter Fatality Reports*, Quincy, Mass.: National Fire Protection Association.

National Fire Protection Association. (1997). *NFPA 1999—Standard on Protective Clothing for Emergency Medical Operations*, Quincy, Mass.: National Fire Protection Association.

National Fire Protection Association. (2000). *NFPA 1971—Standard on Protective Ensemble for Structural Fire Fighting*, Quincy, Mass.: National Fire Protection Association.

National Fire Protection Association. (2001a). *NFPA 1851—Standard on Selection, Care, and Maintenance of Structural Fire Fighting Protective Ensembles*, Quincy, Mass.: National Fire Protection Association.

National Fire Protection Association. (2001b). *NFPA 1994—Standard on Protective Ensembles for Chemical/Biological Terrorism Incidents*, Quincy, Mass.: National Fire Protection Association.

National Fire Protection Association. (2001c). *NFPA 1951—Standard on Protective Ensemble for USAR Operations*, Quincy, Mass.: National Fire Protection Association.

National Fire Protection Association. (2002a). *Fire Department Calls*, Quincy, Mass.: National Fire Protection Association (available at http://www.nfpa.org/PDF/OS.fdcalls.PDF; last accessed November 2002).

National Fire Protection Association. (2002b). *NFPA 1981—Standard on Open-Circuit Self-Contained Breathing Apparatus for Firefighters*, Quincy, Mass.: National Fire Protection Association.

National Institute of Justice. (2002). *Guide for the Selection of Personal Protection Equipment for Emergency First Responders*, NIJ Guide 102–00, Vols. I, IIa, IIb, and IIc, Washington, D.C.: National Institute of Justice (available at http://www.ojp.usdoj.gov/nij/pubs-sum/191518.htm; last accessed December 2002).

National Law Enforcement Officers Memorial Fund. (2002a). *Year By Year Deaths*, Washington, D.C.: National Law Enforcement Officers Memorial Fund (available at http://nleomf.org/FactsFigures/yeardeaths.htm; last accessed August 2002).

National Law Enforcement Officers Memorial Fund. (2002b). *Law Enforcement Officer Line of Duty Deaths, 1992–2001*, database report, Washington, D.C.: National Law Enforcement Officers Memorial Fund (obtained via personal communication from Berneta Spence, April 2002).

National Public Safety Information Bureau. (2002). *National Directory of Fire Chiefs and EMS Administrators 2002*, Stevens Point, Wis.: National Public Safety Information Bureau.

Panlilio, Adelisa. (March 2002). National Surveillance System for Health Care Workers, personal communication.

Peterson, D. J., Tom LaTourrette, and James T. Bartis. (2001). *New Forces at Work in Mining: Industry Views of Critical Technologies*, Santa Monica, Calif.: RAND, MR-1324-OSTP (available at http://www.rand.org/publications/MR/MR1324/).

Reaves, Brian A. (1992). *State and Local Police Departments, 1990*, Washington, D.C.: Bureau of Justice Statistics.

Reaves, Brian A. (1994). *Federal Law Enforcement Officers, 1993*, Washington, D.C.: Bureau of Justice Statistics.

Reaves, Brian A., and Timothy C. Hart. (2001). *Federal Law Enforcement Officers, 2000*, Washington, D.C.: Bureau of Justice Statistics.

Reaves, Brian A., and Matthew J. Hickman. (2002). *Census of State and Local Law Enforcement Agencies, 2000*, Washington, D.C.: Bureau of Justice Statistics.

Runnels, Victoria. (March 2003). International Association of Fire Chiefs, personal communication.

Schwabe, William, Lois M. Davis, and Brian A. Jackson. (2001). *Challenges and Choices for Crime-Fighting Technology: Federal Support of State and Local Law Enforcement*, Santa Monica, Calif.: RAND, MR-1349-OSTP/NIJ (available at http://www.rand.org/publications/MR/MR1349/).

"State and Province Survey." (2001). *Emergency Medical Services*, Vol. 29, No. 12, pp. 207–239.

U.S. Fire Administration. (1998). *National Fire Incident Reporting System Database*, Firefighter Casualty Module, Emmitsburg, Md.: U.S. Fire Administration.

U.S. Fire Administration. (2002). *USFA Releases Preliminary Firefighter Fatality Statistics for 2001*, Emmitsburg, Md.: U.S. Fire Administration (available at http://www.usfa.fema.gov/dhtml/media/02-004.cfm; last accessed December 2002).

U.S. Fire Administration and National Fire Protection Association. (2002). *A Needs Assessment of the U.S. Fire Service*, Emmitsburg, Md.: U.S. Fire Administration.